Macro–ion Characterization

ACS SYMPOSIUM SERIES **548**

Macro–ion Characterization

From Dilute Solutions to Complex Fluids

Kenneth S. Schmitz, EDITOR

University of Missouri—Kansas City

Developed from a symposium sponsored
by the Divisions of Polymer Chemistry, Inc.,
and Colloid and Surface Chemistry
at the 205th National Meeting
of the American Chemical Society,
Denver, Colorado,
March 28–April 2, 1993

American Chemical Society, Washington, DC 1994

Library of Congress Cataloging-in-Publication Data

Macro-ion characterization: from dilute solutions to complex fluids /
Kenneth S. Schmitz, editor.

p.   cm.—(ACS symposium series, ISSN 0097–6156; 548)

"Developed from a symposium sponsored by the divisions of Polymer
Chemistry, Inc., and Colloid and Surface Chemistry at the 205th
National Meeting of the American Chemical Society, Denver, Colorado,
March 28–April 2, 1993."

Includes bibliographical references and index.

ISBN 0–8412–2770–5

1. Polyelectrolytes—Congresses.

I. Schmitz, Kenneth S., 1943–   . II. American Chemical Society.
Division of Polymer Chemistry. III. American Chemical Society.
Division of Colloid and Surface Chemistry. IV. Series.

QD382.P64M33   1994
547.7′04572—dc20   93–39732
                                                         CIP

Foreword

THE ACS SYMPOSIUM SERIES was first published in 1974 to provide a mechanism for publishing symposia quickly in book form. The purpose of this series is to publish comprehensive books developed from symposia, which are usually "snapshots in time" of the current research being done on a topic, plus some review material on the topic. For this reason, it is necessary that the papers be published as quickly as possible.

Before a symposium-based book is put under contract, the proposed table of contents is reviewed for appropriateness to the topic and for comprehensiveness of the collection. Some papers are excluded at this point, and others are added to round out the scope of the volume. In addition, a draft of each paper is peer-reviewed prior to final acceptance or rejection. This anonymous review process is supervised by the organizer(s) of the symposium, who become the editor(s) of the book. The authors then revise their papers according to the recommendations of both the reviewers and the editors, prepare camera-ready copy, and submit the final papers to the editors, who check that all necessary revisions have been made.

As a rule, only original research papers and original review papers are included in the volumes. Verbatim reproductions of previously published papers are not accepted.

M. Joan Comstock
Series Editor

Contents

SOLUTION STRUCTURE

SURFACE ADSORPTION AND INTERFACES

GELS

INDEXES

Preface

POLYELECTROLYTES remain one of the least-understood states of condensed matter, in contrast to neutral polymer solutions. Although bulk properties like osmotic pressure and viscosity have been studied for many years, an understanding of the behavior of polyelectrolytes is lacking. This problem is especially critical because of the fundamental importance of many of the prototypical polyelectrolytes such as, for example, DNA.

Few meetings have focused on polyelectrolytes since the NATO meeting two decades ago. Consequently, the response to the symposium on which this book is based was overwhelming. The participants represented a scientific diversity that covered both theory and experiment, from synthesis to physical properties in (and of) gels, and from nonperturbative methods to the application of external electric fields. The backgrounds of the participants spanned academia and industry; synthetic polymers and biopolymers; and the departments of physics, chemistry, polymers, material sciences, and biological sciences. The participants came from all parts of the world.

The chapters are divided into six major subject areas: theory, synthesis and characterization, condensation and complexation, solution structure, surface adsorption and interfaces, and gels. The objective in making these classifications was to proceed from the more fundamental and simplest cases to the most complex system, with respect both to the organization of the major headings and also to the ordering of the chapters within these sections. The chapters describe a wide variety of techniques as applied to the understanding of these complex systems.

Acknowledgments

Paul Dubin and Ray Farinato put in a tremendous effort in originating and organizing the symposium on which this book is based. I also acknowledge the cooperation of everyone involved in bringing this volume into existence, including the authors, the reviewers, and the people at the American Chemical Society Books Department. I give special thanks to Anne Wilson, whose electronic-mail messages were inspirational. I also thank Julie Fisher for designing and drawing the roller coaster illustration in Chapter 1. Finally, I acknowledge those who were also influential in the organization of this volume: Candace, wherever she may be; Julie Fisher, for her courage; and DAX, may his principles survive.

KENNETH S. SCHMITZ
Department of Chemistry
University of Missouri
Kansas City, MO 64110

August 6, 1993

Chapter 1

An Overview of Polyelectrolytes

Six Topics in Search of an Author

Kenneth S. Schmitz

Department of Chemistry, University of Missouri—Kansas City, Kansas City, MO 64110

Interests in solutions and suspension of macroions has been renewed in the past few years, as evidenced by the mini symposium upon which this present volume is based. Presented in this introductory chapter is a brief presentation of a selection of six topics that I believe deserves further consideration regarding these complex systems: 1) the definition of the screening parameter; 2) the mathematical form of the pairwise electrostatic energy between macroions; 3) the phenomenon of counterion condensation and its application to specific molecular geometry; 4) the use of the Gibbs versus the Helmholtz free energy in describing the thermodynamics of these charged systems; 5) the calculation of the electrostatic component to the persistence length of flexible polyions; and 6) the splitting of relaxation modes as detected by dynamic light scattering measurements.

Studying polyelectrolytes is like taking a roller coaster ride. In the beginning one knows nothing about the field and therefore faces a steep upward learning curve. At some point one reaches a degree of understanding of the field and begins to publish the new results. Further investigations into the field have their ups and downs, sometimes even being thrown for a loop when unexpected results are obtained. Eventually, however, the excursion through the field of polyelectrolytes returns one to the starting point. To illustrate the above metaphor one only has to follow the history of a simple basic question: "Is the interaction between the charged macroions attractive or repulsive?"

In 1938 Langmuir treated a solution of micelles in a manner similar to that of a simple salt crystal. He thus considered the counterions between the micelles as part of the interaction energy, which led to a major result (*1*):

> The interaction of these charges, just as in the sodium chloride crystal, gives an excess of attractive force and in order to have equilibrium it will be necessary not to have an additional *attractive* force, such as the van der Waals force postulated by Hamaker, but some new kind of *repulsive* force.

In contrast to the conclusion of Langmuir, the currently accepted paradigm for the majority of the practitioners in the colloid field is the DLVO (Derjaguin-Landau-

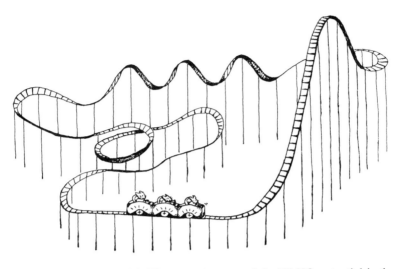

Verwey-Overbeek) model (2). The attractive part of the DLVO potential is the van der Waals contribution that becomes significant at relatively close distances, viz., less than 20 Å. The <u>repulsive</u> part of the DLVO potential is a <u>pairwise screened</u> <u>Coulomb</u> interaction between two colloidal particles of radius R_S and charge $q_p = Z_p q_e$ (q_e is the magnitude of the electron charge) separated by a distance R is,

$$U_{elec}^{DLVO} = \left(\frac{e^{\kappa_{DH}R_S}}{1+\kappa_{DH}R_S}\right)^2 \frac{q_p^2}{\varepsilon} \frac{e^{-\kappa_{DH}R}}{R} = X_{DLVO}^2 \frac{q_p^2}{\varepsilon}\frac{e^{-\kappa_{DH}R}}{R} \qquad (1)$$

where the Debye-Hückel screening parameter κ_{DH} is defined by the m_{add} different species of <u>added</u> electrolyte ions at concentrations $<n_j>_u$ [uniform (bulk) concentration in particles/mL] and charge $Z_j q_e$,

$$\kappa_{DH}^2 = \frac{4\pi q_e^2}{\varepsilon k_B T}\sum_{j=1}^{m_{add}} Z_j^2\langle n_j\rangle_u = 4\pi\lambda_B \sum_{j=1}^{m_{add}} Z_j^2\langle n_j\rangle_u \qquad (2)$$

where ε is the dielectric constant, k_B is Boltzmann's constant, and λ_B is the Bjerrum length. Note that only the <u>added</u> electrolyte ions contribute to the screening parameter in the DLVO theory as in the Debye-Hückel theory of simple salts, which is why the subscript "DH" is include in equation 2. The reason for the apparent universal acceptance of the DLVO potential lies in its qualitative success in explaining most phenomena in colloidal systems even though the apparent charge required to fit the data is generally <u>smaller</u> than the anticipated value of the charge based on the chemical composition of the macroion.

For more than two decades, however, Ise and co-workers (3-10) have emphasized that there exists some form of long range *attractive* interaction between macroions that is not consistent with the standard DLVO potential. More specifically, video movies taken through an ultramicroscope clearly showed two distinct regions in a suspension of latex particles; a highly kinetic region in which the particles undergo Brownian motion, and a relatively static ordered region (5, 9) which has embedded regions of stable voids (9). Sogami and Ise (11) thus developed a model based on the Gibbs free energy that exhibited a stable minimum at macroion separation distances on the order of several thousand angstroms, as experimentally observed. Overbeek (*12*), however, criticized this model on the basis that the electrical contribution of the solvent exactly canceled the term $\kappa_{DH}^2(\partial A/\partial\kappa_{DH}^2)_{V,T,Z}$ that gave rise to the minimum in the Sogami-Ise potential, where A is the Helmholtz free energy. From the Gibbs-Duhem relationship Overbeek (*12*) derived the solvent contribution,

$$\langle n_o\rangle_u\left[\frac{\partial A}{\partial\langle n_o\rangle_u}\right]_{V,T,Z} = -\kappa_{DH}^2\left[\frac{\partial A}{\partial\kappa_{DH}^2}\right]_{V,T,Z} \tag{3}$$

where $\langle n_o\rangle_u$ is the uniform distribution of the solvent. Overbeek thus concluded:

> The claim in Sogami's and Ise's papers that G^{el} leads to an attraction between particles is incorrect. The experimental data presented in papers 1 through 7 and in other similar publications will have to be explained in a different manner.

This conclusion is thus diametrically opposite of the conclusion of Langmuir as to whether an additional attractive or repulsive force, respectively, must be postulated to explain the stability of colloidal systems.

The story does not end here, however. Ise and co-workers (*10*) point out in a footnote that Overbeek misapplied the Gibbs-Duhem expression by *not also including the macroion contribution*. Citing Verwey and Overbeek (*2*) in regard to the specific terms that comprise the electrical contribution tin the DLVO theory, Smalley (*13*) points out that the Overbeek solvent correction likewise cancels the electrical contribution to the DLVO potential. Smalley thus concluded that in the DLVO theory with the solvent term (*13*):

> there is *no* energy associated with the electrical double layers.

Since experimental data clearly establishes that there <u>are</u> electrical interactions between charged particles in solution and colloidal suspension one must conclude from the <u>symmetric</u> application of the solvent contribution term of Overbeek to both the DLVO and SI potentials that either the DLVO theory is wrong or that the nature of the electrical interaction between charge macroions under the most dilute conditions is not well understood.

It may be suggested from the previous paragraph that we appear to have returned to the beginning of the roller coaster ride, i.e., knowing nothing about the fundamental interactions between macroions in these complex systems. This statement is not entirely true, however, since certain concepts have been developed towards the understanding the basic nature of macroionic systems. The illusion of truth in that sentence, however, is manifested in the literature by varying usage of these concepts in the description of the asymmetrically charged systems of polyions and simple salts. The following few paragraphs address some of these problems, with the intent to stimulate further discussion on these matters and not to resolve their differences.

The Screening Parameter

The most fundamental concept in systems of charged particles is the screening parameter, generically denoted by the symbol κ. In the Debye-Hückel theory of simple electrolytes κ is defined in terms of <u>all</u> of the electrolyte species present since only one of the ions is selected as the "test" ion.

Definition of the Screening Parameter in Macroionic Systems. In macroionic systems the contribution of the macroions and associated counterions are not included in the DLVO theory, where the screening length as defined by equation 2. The justification for the omission of the macroion and associated counterion contributions is that the concentration of these components is sufficiently small that they can be ignored. This result is clearly indicated in the mean spherical approximation (MAS) to the Ornstein-Zernike equation, as shown for example by Median-Noyola and McQuarrie (*14*). In the coupled mode theory of Lin, Lee, and Schurr (*15*), however, the total screening parameter naturally results from the solution to the cubic secular equation for the polyion-added electrolyte system, viz.,

$$\kappa_{\text{tot}}^2 = 4\pi\lambda_B\left(Z_1^2\langle n_1\rangle_u + Z_2^2\langle n_2\rangle_u + Z_3^2\langle n_3\rangle_u\right) \tag{4}$$

where the polyion, counterion, and coion are denoted by 1, 2, and 3, respectively, and the electroneutrality constraint is employed,

$$\langle n_2\rangle_u = -\frac{Z_1\langle n_1\rangle_u + Z_3\langle n_3\rangle_u}{Z_2} \tag{5}$$

There is a third definition of the screening parameter that is intermediate between those defined by equations 2 and 4 and employed, for example, by Beresford-Smith and coworkers (*16*) and by Belloni (*17*). In this definition the counterions from the macroion are included in the concentration of the common ion from the added electrolyte but the contribution of the macroion is omitted, viz.,

$$\kappa_{\text{interm}}^2 = 4\pi\lambda_B\left[Z_2^2\left(\langle n_{2,\text{add}}\rangle_u + |Z_1|\langle n_1\rangle_u\right) + Z_3^2\langle n_3\rangle_u\right] \tag{6}$$

The consequences of the choice of the screening parameter are discussed in more detail in the section of interaction potentials.

"Global" and "Local" Screening Parameters. The screening parameters defined by equations 2, 4, and 6 may be referred to as "global" values since they are a function of the uniform concentrations of the ions. The electrolyte ions, however, are not uniformly distributed throughout the solution. The distribution of electrolyte ions about a polyion, however, is strongly dependent upon the magnitude and distribution of surface charge of the polyion. Counterions are preferentially drawn to the vicinity of the polyion whereas coions are repelled from this region. For the Boltzmann distribution is easily shown that in the limit of weak polyion charge that the excess number of counterions exactly cancels the deficit number of coions and the screening parameter (for symmetric electrolyte ions) truly takes on a "global" character.

In their cell model for the electrolyte ion distribution about a charged cylinder, Alexandrowicz and Katchalsky (*18*) partitioned the problem into two regions. The "inner region" was assumed to contain only the counterions since it was presumed that the high charge density prohibited the coions from accumulating in significant concentrations. The solution to the Poisson equation thus employed only the local concentration of counterions at this boundary. Because of symmetry the upper limit of the "outer region" was defined at the mid-distance between the aligned rods. The linearized Debye-Hückel solution to the Poisson equation was assumed to be valid in this region, where the counterion concentration was necessarily larger than the coion concentration. Russel (*19*) pointed out that the parallel aligned cylinders in the Alexandrowicz and Katchalsky calculation was not realistic, and that at the outer boundary should be defined by the thickness of the ion cloud. Hence the periodic boundary conditions were replaced with a random distribution of neighboring macroions. The Alexandrowicz - Katchalsky model in the cylindrical system therefore set a precedence to the assumption of "local screening parameters" in the Sogami-Ise (*11*) and Sogami (*20*) theories for the interaction between charged spheres. It has also been proposed that the local screening parameter for the interaction between discrete charges on a cylinder be computed as a "linear average" of the relative concentrations of "inner" (sheath) and "outer" (bulk) counterion concentrations (*21*).

Comments on Screened Electrical Interactions.   There is a universal agreement that electrical interactions between charged particles are screened. The screening of interactions between two macroions originates from the principle that the electrolyte ions are mobile in the solution or suspension. Because of the complexity of the many-body problem, however, the computations generally involve only three particles, viz., the two particles of interest at a fixed separation distance while the integration is carried out over the coordinates of a mobile third particle. Hence both the correlations between the mobile particles and the contributions of the other macroions are neglected. A more focused study of electrolyte correlations and distributions about the macroions and the resulting effects on the macroion-macroion interactions should be initiated. These results could then be compared with the results of the *ad hoc* operational definitions of the screening parameter.

Pairwise Electrostatic Energy Between Macroions

Introduced in this section are selected forms of the pairwise electrostatic interaction potential between two identical macroions. Those given in the first subsection are based on the electrostatic Helmholtz free energy, and the second subsection presents the Sogami-Ise result (*11*) based on the Gibbs free energy. The discussion regarding the use of the Helmholtz or the Gibbs free energy depends upon concepts related to counterion condensation given in the following section. Hence the Helmholtz versus Gibbs free energy is postponed until then.

Helmholtz Free Energy.   Since the screening parameter in the DLVO potential is dependent only on the concentration of <u>added</u> electrolyte and independent of the macroion charge and concentration, the DLVO potential is said to be a "true" pairwise interaction potential. That is, the value of U_{elec}^{DLVO} is independent of the polyion concentration. In contrast other pairwise interaction potentials presented herein are dependent upon the polyion concentration.

Using the "Jellum Approximation" of uniformly distributed spherical macroions, Beresford-Smith, Chan, and Mitchell (*16*) obtained the pairwise interaction potential,

$$U_{elec}^{BSCM} = \left[\frac{e^y (1+\phi_1)}{1+y} \right]^2 \frac{q_1^2}{\varepsilon} \frac{e^{-\kappa_{interm}R}}{R} = X_{BSCM}^2 \frac{q_1^2}{\varepsilon} \frac{e^{-\kappa_{interm}R}}{R} \qquad (7)$$

where ϕ_1 is the volume fraction of the polyion, $y = \kappa_{interm}R_S$, and κ_{interm} is defined by equation 6. Note that <u>at fixed R</u> increasing the concentration of the polyion the potential in equation 7 becomes more repulsive than in the DLVO potential due, primarily, to the differences in the definition of the screening parameter.

Belloni (*17*) pointed out that the Jellum Approximation was not needed since the MSA solution to the Ornstein-Zernike equation yielded an exact solution for the hard sphere potential. Belloni obtained for the pairwise interaction potential,

$$U_{elec}^{B} = X_B^2 \frac{q_1^2}{\varepsilon} \frac{e^{-\kappa_{interm}R}}{R} \qquad (8)$$

where

$$X_B = cosh(y) + [y\,cosh(y) - sinh(y)]\left[\frac{3\phi_1}{y^3(1-\phi_1)} - \frac{\gamma}{y}\right] \tag{9}$$

and γ is a "screening function" that depends upon ϕ_1 and the MSA Coulombic screening parameter whose precise form is of no consequence in this discussion.

Sogami and Ise (*11*) approached the macroion-electrolyte ion problem on the same footing as the quantum mechanical description of chemical bonding. Using the adiabatic approximation for separation of macroion and electrolyte ion contributions, they expressed the charge density in the Poisson equation as two parts, that which was due to the macroions and that which was due to the electrolyte ions. Hence there was an interdependence of the distribution of both the electrolyte ions and the macroions. They then employed the usual charging procedure as in the DLVO theory and obtained the resulting expression for the Helmholtz interaction potential,

$$U_{elec}^{SI} = \left[\frac{sinh(y)}{y}\right]^2 \frac{q_1^2}{\varepsilon}\frac{e^{-\kappa_{interm}R}}{R} = X_{SI}^2\frac{q_1^2}{\varepsilon}\frac{e^{-\kappa_{interm}R}}{R} \tag{10}$$

Sogami (*20*) proposed a model based on the linearized Debye-Hückel potential of interaction between two charged spheres with a two-state electrolyte ion distribution. Region I was defined in terms of the macroion and the associated ion cloud of thickness λ_C. The effective radius of the kinetic unit, R_S^*, is therefore $R_S + \lambda_C$. Region II was the solution intermittent to the two kinetic units. Since the kinetic unit was assumed to be charged it followed that for region II,

$$\frac{N_2^{II}q_2 + N_3^{II}q_3}{V^{II}} \neq 0 \tag{11}$$

The value of the screening parameter κ_{II} was then computed from the number concentrations in equation 11. Sogami then obtained for the interaction potential,

$$U_{elec}^S = X_S^2\frac{q_1^2}{\varepsilon}\frac{e^{-\kappa_{II}R}}{R} \tag{12}$$

where

$$X_S = \frac{q_1^*}{q_1}\left\{1 - \frac{q_1 - q_1^*}{q_1^*}\left[2 - \frac{1}{4}\left(\frac{N_2^{II} - N_3^{II}}{N_2^{II} + N_3^{II}}\right)\left(e^{-2\kappa_{II}R_S^*} - 1 + 2\kappa_{II}R_S^* - \kappa_{II}R\right)\right]\right\}^{1/2} \tag{13}$$

and

$$q_1^* = \frac{sinh(\kappa_{II}R_S^*)}{\kappa_{II}R_S^*}\left(q_1 + N_2^I q_2 + N_3^I q_3\right) \tag{14}$$

All of the above forms of the Helmholtz electrostatic interaction potential have the Yukawa form [i.e., exp(-aR)/R] multiplied by a model specific form X^2. At this stage of discussion only the Sogami model for the Helmholtz electrical interaction potential has the property of exhibiting a minimum. The reason for this minimum is found in the expression for the location of the minimum (*20*),

$$R_{min} = \frac{4\left(\dfrac{N_2^{II}+N_3^{II}}{N_2^{II}-N_3^{II}}\right)\left(\dfrac{3q_1^*-2q_1}{q_1-q_1^*}\right)+e^{-2\kappa_{II}R_S^*}+2\kappa_{II}R_S^*}{\kappa_{II}} \tag{15}$$

It now becomes transparent that the presence of a minimum at finite separation distances is a result of a difference in the counterion and coion concentration in the Region II. Hence the situation is reminiscent of, but not identical to, the Alexandrowicz and Katchalsky (*18*) model discussed in the previous section. By setting $N_2^{II} = N_3^{II}$ the Sogami potential reduces to the same form of the DLVO potential with the location of the minimum at infinity. This transformation is thus similar in concept to the transformation from the Alexandrowicz-Katchalsky model (*18*) to the model of Russel (*19*) for charged cylinders.

Gibbs Free Energy. Sogami and Ise (*11*) proposed a theory of the interaction potential between two macroions based on the Gibbs free energy. The interaction potential is obtained from equation 10 using the operator $2 + \kappa^2\left(\partial/\partial\kappa^2\right)$, viz.,

$$G_{elec}^{SI} = U_{elec}^{SI}\left[1 + y\,coth(y) - \kappa_{interm}R\right] \tag{16}$$

where U_{elec}^{SI} is defined by equation 10. A minimum in G_{elec}^{SI} occurs at (*11*),

$$R_{min} = \frac{y\,coth(y)+1+\left\{\left[y\,coth(y)+1\right]\left[y\,coth(y)+3\right]\right\}^{1/2}}{\kappa_{interm}} \tag{17}$$

Sogami and Ise (*11*) estimated that $R_{min} \approx 5.47 R_S$.

Comments on the Pairwise Interaction Potential. The DLVO potential is a "true" pairwise potential since it is independent of the concentration of the macroions. The remaining potentials discussed in this section reduce to the DLVO potential in the limit of vanishing macroion concentration and uniform electrolyte concentration. Hence the differences in these potentials arise from the *ad hoc* assumptions regarding the disposition of the macroion contributions at finite concentration and their effect on the electrolyte ion distribution. Regardless of the choice of potential used in the analysis of experimental data, a common result is that the apparent charge of the macroion is considerably smaller than that anticipated on the basis of chemical analysis and titration curves. This discrepancy in charge values has not been vigorously addressed in the literature, but partially "justified" by postulating "charge renormalization" concepts or redefining the structural unit as the macroion and tightly associated counterions.

The primary problem with the aforementioned potentials is that they are mean potentials developed from the Debye-Hückel limiting expressions. The situation with macroion systems is therefore reminiscent of the Faraday discussion in 1927 regarding deviations of experimental data on the simple salt systems from the predictions of the Debye-Hückel theory. For example, the Debye-Hückel potential obtained from the linearized Poisson-Boltzmann equation is consistent with the fundamental principles of the Gibbs phase integral in statistical mechanics as long as fluctuations in the mean potential of simple electrolyte systems are negligible (*22*). The possibility arises that fluctuations in the mean potentials are of even greater significance for the highly asymmetric system of macroions and electrolyte ions.

Counterion condensation: the Oosawa-Manning theories

Oosawa and co-workers (*23-27*) employed the Poisson-Boltzmann equation to study the interaction of small ions with the geometry of charged spheres, rods, and coils. They proposed two types of small ion association mechanisms; those localized at specific chemical groups on the polyion ("P-binding") and those delocalized within a volume (which we denote as v_c) of the polyion ("ψ-binding") (*24*). A major conclusion of these studies was that if the charge parameter λ_c was greater than unity, then the average extent of dissociation, $\langle \lambda_d \rangle$, of the cylindrical polyion was,

$$\langle \lambda_d \rangle \ = \ \frac{1}{\lambda_c} \ = \ \frac{b}{|z_1 Z_s| \lambda_B} \qquad (\lambda_c > 1) \tag{18}$$

where z_1 is the charge on the monomer group for the polyion, b is the distance between groups, and Z_s is the charge of the symmetric electrolyte.

Manning (*28-34*) developed a "two-state" model for counterion condensation about an infinitely long line charge. The primary assumption in the Manning model was that the reduction in electrical free energy between the *charged groups on the line charge* due to condensation of counterions was counterbalanced by the entropy of mixing. The sum of these two contribution was then minimized to obtain the critical fraction of bound sites by the counterions, θ_b^c. Because of the linear geometry and infinite chain length assumptions the summation over electrostatic energy states resulted in a logarithmic expression for the electrostatic free energy. In order to eliminate the divergence in the free energy expression as the counterion concentration went to zero, the coefficient of the term $ln(< n_2 >_u)$ was set equal to zero. Hence,

$$|Z_2| \theta_b^c \ = \ 1 - \frac{b}{|Z_2| \lambda_B} \tag{19}$$

The relationship between the Oosawa and Manning models is,

$$\langle \lambda_d \rangle \ = \ 1 - |Z_s| \theta_b^c \ = \ \frac{1}{|Z_s| \xi_{OM}} \tag{20}$$

where the Oosawa-Manning parameter ξ_{OM} is defined as,

$$\xi_{OM} \ \equiv \ \frac{\lambda_B}{b} \tag{21}$$

The physical meaning of the Oosawa-Manning parameter is that it represents the maximum charge density that can be supported by the solvent. If $\xi_{OM} < 1$ then the average spacing between the charges on the rodlike polyion is simply the charge spacing as determined, for example, by titration and <b> = b. On the other hand if $\xi_{OM} > 1$ then counterions will condense onto the linear surface until the condition $\xi_{OM} = 1$ is met, i.e., <b> = λ_B. The same conclusion results if the nonlinear Poisson-Boltzmann equation is solved (*19*).

The Oosawa and Manning theories are reminiscent of the Bjerrum model for simple salts. Those counterions that are "condensed" onto the polyion surface are "thermodynamically removed" from the bulk solution and also effectively reduce the charge of the polyion. Caution must be exercised, however, in the application of the

Oosawa-Manning condensation results to linear polyions of finite length and thickness, to polyions with geometric shapes other than the line charge, to site specific binding systems, and to ligands of finite size. The assumption of a line charge, for example, appears to be valid for extremely low ionic strength solvents (large λ_{DH}) but fails at moderate to high ionic strength solvents (*35-37*). Skolnick and Grimmelmann (*38*) show that in the infinite length approximation may be valid for finite chain length only if λ_{DH} is small relative to the length of the line. Using the Mayer cluster integrals Woodbury and Ramanathan (*39*) conclude that *counterions do not condense onto finite length line charges* . If the dielectric constant of the polyion differs from that of the bulk solvent then projection of the actual charge distribution onto a line may lead to difficulties in the application of the Oosawa-Manning condensation theories. Skolnick and Fixman (*40*) show that charges placed on the same side of a cylinder *enhanced* the site-site interactions relative to the reference condition of two charges in the solvent, whereas the interaction was decreased if the two charges where placed on opposite sides of the cylinder. Dewey (*41*) notes that the "condensation" volume goes to infinity as the electrolyte concentration goes to zero for finite chain lengths because the electrostatic instability for the infinite chain length no longer exists. According to Dewey (*41*),

> Since this instability no longer exists for finite length polymers, retention of the condensation free energy term now causes an instability in the system.

A problem remains in the "operational definition" of the condensed counterions. NMR measurements on the association of $^{23}Na^+$ to DNA indicates that the cation is mobile, and thus "territorially" bound and rapidly exchanges with those in the bulk solution (*42,43*). On the other hand NMR studies indicate that $^{25}Mg^{2+}$ binds in a more complex manner, having an "inner sphere" with an accompanying hydration change and an "outer sphere" with no hydration change (*44, 45*). Although NMR studies indicate that the hydration shell of hexaamine cobalt (III) [CoHex(III)] does not appear to be altered on binding to DNA (*46*), there does appear to be at least three different binding classes for this cation (*47*). It is emphasized that the NMR methods cannot detect those ions "trapped" in the ion cloud, and thus these ions are not counted as being in the "bound" state by the NMR technique. Granot and Kearns (*45*) state, for example,

> those ions which are only held in an ion atmosphere around the polyelectrolyte by long-range, through space, electrostatic interactions, would not be considered bound and are not detected by the PMR method.

This definition of "condensed" counterions is not universally applied to all techniques used to study counterion condensation. Electrophoretic mobilities, however, are sensitive to electroviscous effects whose origins are associated with the thickness of the surrounding ion cloud of the polyion. Electrophoretic light scattering (ELS) methods therefore cannot distinguish between counterions tightly bound to the surface of the macroion and those that are trapped in the ion cloud. Nonetheless ELS methods are also interpreted in terms of counterion condensation (*48, 49*). In the study of Klein and Ware (*49*) on 6,6-ionene, for example, a sharp change in the electrophoretic mobility occurs at $\xi_{OM} = 1$ over the range $0.82 < \xi_{OM} < 1.82$ in accordance with the Oosawa-Manning prediction.

Of direct bearing to the question of the condensation volume are the results of Le Bret and Zimm (*50*). The computational model used by these authors is a composite of the Alexandrowicz and Katchalsky (*18*) and Russel (*19*) models for the Poisson-Boltzmann equation. In these calculations the "salt free" expressions are used in the vicinity of the charged cylindrical surface while the "excess salt" expressions are used far from the surface but with the constraint that the screening parameter is the same in both regions. The "Manning distance" R_M is thus defined as that distance from the charged surface such that the inscribed volume contains the number of electrolyte ions predicted by the Manning condensation expression. From this analysis they conclude that R_M is proportional to $\kappa^{-1/2}$, and thus increases as the electrolyte concentration decreases in value. Hence the Le Bret-Zimm calculations are more consistent with the idea that "condensed" counterions are those trapped in the vicinity of the polyion, including those in the distal regions of the ion cloud as well as the tightly bound as detected by NMR.

Comments on the Concept of Counterion Condensation. There is ample experimental evidence that supports the concept of counterion condensation. The "two-state" models introduced by Oosawa and by Manning provide a very good quantitative picture of counterion condensation for very long polyions whose linear charge density is accurately represented as a line charge. Conspicuously missing in the Manning theory is any reference to the electrostatic interactions of the electrolyte ions. For example, there is no term that reflects the association energy of the electrolyte ion with the charged line. Trapping a counterion within the volume v_c serves only to <u>uniformly</u> reduce the repulsive interactions <u>between the lattice sites</u>. Likewise there is no term for the pairwise interaction between "bound" counterions within the "condensation" volume, v_c. The latter contribution does not appear for point charges that are "absorbed" into the lattice charge, i.e., the coincidence of the centers of positive and negative charge. Ions of finite size, however, do not result in complete neutralization of the site charges but rather give rise to "dipole" lattices (51). The apparent success of the Manning formulation of counterion condensation to large DNA fragments may be due to the observation that ion binding to DNA is largely entropy driven (see the paper by Bloomfield, Ma, and Arscott in this volume) or that experimental data is over a relatively narrow concentration range of the ligand (51). Further tests on the counterion condensation models should be carried out on systems of very flexible polyions and/or enthalpy driven associations of the ions.

Gibbs Versus Helmholtz Free Energy in Describing Macroion Systems

Both the Oosawa and Manning models employ the somewhat ill-defined quantity of the "condensation volume" that contains "thermodynamically bound" counterions. That is, the "bound" counterions act to reduce both the net charge of the macroion and the effective concentration of counterions in the bulk solution. If the "condensation volume" depends upon the ionic strength of the solution as indicated from the Le Bret-Zimm calculations (50), then it necessarily follows that the "free volume" is likewise a variable quantity. Thus the question is now raised as to whether the Helmholtz or the Gibbs free energy is the appropriate function to describe the thermodynamics of macroionic solutions and suspensions.

Matsumoto and Kataoka (52) point out that the volume of the solution accessible to the small ions ($V_{f,s}$) differs from that of the polyions ($V_{f,1}$). To illustrate this concept we note that both the Oosawa and the Manning models partition the solution that is accessible <u>to the small ions</u> into two volumes; the condensation volume about the polyion containing the thermodynamically "bound" counterions (v_c), and the remainder of the solution containing the thermodynamically "free" electrolyte ions ($V_{f,s}$). These two quantities are related to the total volume of the solution, V, by the expression,

$$V_{f,s} = V - N_1 v_c \tag{22}$$

where N_1 is the number of macroions present in the volume V. This equation is similar to that employed by Oosawa, Imai, and Kagawa (*23*), except that now v_c can vary with both the charge of the macroion and the concentration of charged molecules in accordance with the Le Bret-Zimm calculations of R_M (*50*). The net effect is that the free volume accessible to the electrolyte ions is a variable quantity. In contrast the total volume accessible to the polyion is dependent upon the physical volume v_1 of the polyion,

$$V_{f,1} = V - N_1 v_1 \qquad\qquad (23)$$

Note that v_1 does not in general vary with the charge or the ionic strength of the solution.

The question of whether the Gibbs or the Helmholtz free energy should be used thus becomes one of whether one should use for the total volume V or $V_{f,s}$ to describe the properties of the "thermodynamically free" electrolyte ions.

Electrolyte Ions and Attraction Between Identical Macroions

As mentioned in the introductory paragraph, Langmuir (*1*) concluded that there is a net attractive electrostatic interaction when the counterions in a non uniform distribution (i.e., between the micelles) were included in the calculation. Kirkwood and Schumaker (*53*) showed that fluctuations in the net charge of spheres resulted in an attraction between the macroions. Oosawa (*54*) showed that an attraction occurs between two parallel rods at distances on the order of the Bjerrum length if the fluctuations in the axial distribution of counterions are correlated. Along similar lines Fulton (*55, 56*) pointed out that fluctuations in the dipole moment of polyions resulted in long range correlations between the charged spheres. Along these lines it was suggested that *distortions* in the electrolyte atmosphere about nearby polyions (*57, 58*), or the *sharing* of electrolyte ions in overlapping ion clouds of neighboring polyions (*59*) are responsible for the occurrence of the very slow relaxation mode reported in dynamic light scattering data. Yoshino (*60*) attempted to explain the "two state" observations of PSLS with no added salt in terms of a distortion from spherical symmetry of the Debye cloud about a charged sphere. Yoshino proposed that these distortions could be represented in terms of multipole contributions as described by spherical harmonics. Recall also that both the Sogami and the Sogami-Ise potentials (equations 12 and 16, respectively) differ from the standard DLVO potential (equation 1) solely on the assumptions regarding the distribution of the electrolyte ions. If the electrolyte ions are uniformly distributed in the medium then equations 1, 12, and 16 become identical in form. In his lecture at this mini symposium, Manning (*61*) presented preliminary results on the interaction between parallel line charges. As the parallel line charges were brought closer together, it was found that the condensation volumes v_c became distorted such that there was an increase in the volume between the line charges. Under certain conditions the free energy associated with the distortions in v_c overcame the repulsive Coulombic contributions with the result of a net attraction between the line charges. Sánchez-Sánchez and Lozada-Cassou (*62*) reported three-point hypernetted chain/mean spherical results for the interaction between two charged spheres. They found that for the higher surface charge densities the counterion distribution between the two spheres was greatly enhanced over that of the bulk solvent. The force between the two spheres was reduced due to the accumulation of intermittent counterions, but no attraction between the two spheres was reported.

Comments on the Role of Electrolyte Ions and Polyion Attraction. A common theme of the studies reviewed in the previous paragraph is the increase in the concentration of counterions between the two charge macroions. These studies differ, however, in the mathematical models employed to describe the interacting systems. The quantitative expression of the Sogami-Ise and Sogami potentials predict a minimum in the potential located 1000-2000Å from the central macroion. Manning reported that a minimum in the net attraction curve appears at 50-60Å separation when distortion in the condensation volume dominates the electrostatic interactions.. The calculations of Sánchez-Sánchez and Lozada-Cassou do not indicate any attraction between the spheres. The absence of attraction may be due to the close proximity of the two spheres, being on the order of 5 sphere diameters. Thus the apparent attraction between the macroions results from the distortion of some characteristic property of the model as compared to that quantity for the isolated macroion. Hence it may be of value to treat the collective effect of small ion-polyion interactions within the vocabulary associated with electrons and atoms.

Electrostatic Persistence Length: Stiff Versus Flexible Polyelectrolytes

Odijk (63) and, independently, Skolnick and Fixman (64) presented a theory for the electrostatic component to the persistence length for an infinitely long polymer based on slight deviations from the rigid rod configuration. This model was corrected for finite chain of length L by Odijk and Houwaart (65), with the result,

$$L_{elec}(\kappa L) \; = \; \frac{\lambda_B}{4\langle b\rangle^2 \kappa^2} h(\kappa L) \tag{24}$$

where <b> is the average charge spacing (to account for Oosawa-Manning condensation) and the finite chain length correction factor is,

$$h(\kappa L) \; = \; 1 - \frac{8}{3\kappa L} + \frac{e^{-\kappa L}}{3}\left(\kappa L + 5 + \frac{8}{\kappa L}\right) \tag{25}$$

Hence the prediction for this model is that L_{elec} is proportional to the reciprocal of the salt concentration, $1/C_S$.

The dependence of L_{elec} on C_S does now always exhibit the -1 power law as given by equation 24. Le Bret (66) solved the Poisson-Boltzmann equation numerically for a toroid both with and without a conducting surface. The power law for a conducting surface tended to follow the -1 power law, but for a non conducting surface it varied from -1 for very low ionic strength solvents to -1/4 for C_S on the order of 1 M. Similar results were found by Fixman (67).

In the case of flexible polyelectrolytes, Tricot (68) pointed out that the experimental data supported a power law of -1/2. Barrat and Joanny (69) took into consideration the effect of fluctuations in the chain configuration on the calculation of the electrostatic component to the persistence length. It was found that the Odijk/Skolnick-Fixman equations are correct in the limit of stiff chains ($L_p/L > 1$) but breaks down for very flexible chains ($L_p/L \ll 1$). Their results indicated that for very flexible chains L_{elec} was proportional to $1/\kappa$, or $C_S^{-1/2}$, as experimentally observed for several polyion systems.

Comments on the Electrostatic Component to the Persistence Length.
Exact numerical calculations of the electrostatic persistence length were performed for a continuously bent rod using a screened Coulomb pairwise interaction (21). These calculations indicated that equations 24 and 25 gave very good values for bending angles much larger than those considered in the original derivation, i.e., in

excess of a bending angle of 180°. It was further shown that if these various degrees of bending were <u>averaged</u> over a Boltzmann distribution then the power law of $\langle L_{elec} \rangle$ was considerably less than -1 even though each contributing configuration obeyed the Odijk/Skolnick-Fixman power law, and that $\langle L_{elec} \rangle$ appeared to become independent of C_S at sufficiently small values of κL.

When trying to interpret the experimental observation that L_{elec} is proportional to $C_s^{-1/2}$ because of fluctuations, one must consider two types of averages. In case 1 there is a single molecular weight species and the average is carried out over all possible configurations of these chains, including the continuous, smooth bending polyion as in the Odijk and Skolnick-Fixman models. In case 2 the sample that is polydisperse in regard to the molecular weight, or chain length, distribution but all of the polyions undergo a continuous, smooth bend as in the Odijk and Skolnick-Fixman models (case 2). Apparently both of these cases give rise to an average value of the electrostatic persistent length that is proportional to C_s^{-n}, where n is less than 1.

Slow Modes in the Dynamic Light Scattering Correlation Function

Lin, Lee, and Schurr (*70*) reported rather bizarre behavior of the [NaBr] profile of the apparent diffusion coefficient, D_{app}, of poly(L-lysine) with a degree of polymerization of 955, denoted by (PLL)$_{955}$. The value of D_{app} increased as [NaBr] decreased from 0.5 M, in accordance with the Einstein expression for the mutual diffusion coefficient, D_m,

$$D_m = \frac{1}{f_m}\left(\frac{\partial\pi}{\partial C_1}\right) \quad \alpha \quad \frac{kT}{f_m}\frac{1}{I_{tils}} \tag{26}$$

where f_m is the mutual friction factor that contains indirect hydrodynamic interactions, $(\partial\pi/\partial C_1)$ is the osmotic susceptibility that measures the direct interaction between the polyions and is proportional to the reciprocal of the total intensity, I_{tils}, and C_1 is the polyion concentration. Hence in this region of the salt profile $D_{app} = D_m$, and the polyion is said to exhibit "ordinary" behavior.

At a well-defined value of [NaBr] that was dependent only on the weight concentration of PLL$_{955}$, D_{app} underwent an <u>unexpected</u> <u>catastrophic</u> drop in value that could not be explained in terms of current polyelectrolyte theory (*70*). For the solution that was 1 mg/mL in PLL, D_{app} changed from a value of $\approx 8 \times 10^{-7}$ cm^2/s at [NaBr] $\approx 1.1 \times 10^{-3}$ M to a value of $\approx 4 \times 10^{-8}$ cm^2/s at [NaBr] $\approx 9.1 \times 10^{-4}$ M, where the latter value of D_{app} was much smaller than that calculated for PLL$_{955}$ modeled as a cylinder. Furthermore, they searched with no avail for two relaxation modes on either side of the transition. In their search in the ordinary regime they stated (*70*),

> There seemed to be virtually no sign of the very slow relaxation characteristic of the extraordinary solutions, however.

They also reported that I_{tils} also changed sharply in value over this range (*70*),

> the 1 mg/mL solution exhibited at $\theta = 60^\circ$ a 2.5 fold increase in intensity between 9.1×10^{-4} and 1.1×10^{-3} M NaBr.

The concomitant <u>decrease</u> in D_{app} and <u>decrease</u> in I_{tils} was therefore contrary to the relationship given by equation 26. Thus D_{app} for [NaBr] $< 10^{-3}$ M was not associated with the mutual diffusion coefficient, but rather is a manifestation of an <u>extraordinary</u> behavior of the dynamics of the PLL955 solution. This *very sharp* drop in the value of D_{app} for a 1 mg/mL solution of PLL at 10^{-3} M univalent salt for a has been verified for PLL406 *(71)*, PLL946 *(71)*, PLL2273 *(71)*, PLL616 *(72)*, PLL2500 *(73)*, PLL3800 *(74)*, and PLL952-PLL1380 *(75)*. Representative data are shown in Figure 1, where the location of the ordinary-extraordinary (o-e) transition is indicated by the solid line. The sharpness of the o-e transition for both the PLL and poly(styrene sulfonate) (PSS) is further reflected in the empirical expression of Drifford and Dalbiez *(76)*. Let us define the ratio ρ_{DD} as

$$\rho_{DD} = \frac{C_m \langle b \rangle}{Z_s \lambda_B \sum_j C_j Z_j^2} \tag{27}$$

where C_m is the concentration of <u>monomer</u> units of the polyion. The empirical relationship for the location of the o-e transition is when $\rho_{DD} = 1$. Hence for a fixed electrolyte concentration this condition obtains when the polyion concentration is,

$$C_m^{o\text{-}e} = \frac{Z_s \lambda_B \sum_j C_j Z_j^2}{\langle b \rangle} \tag{28}$$

Equation 28 is referred to as the Drifford-Dalbiez ratio. Although not as dramatic as in D_{app}, small changes in the relative viscosity *(77)*, conductivity *(78)*, and electrophoretic mobility *(71, 72)* were reported to occur at $C_m^{o\text{-}e}$. In contrast the tracer diffusion coefficient obtained by fluorescence recovery after photobleaching methods did not show any unusual behavior through the o-e transition region *(72)*.

Classification of D_{app} Versus C_s Profiles. Since the original studies on PLL and PSS that <u>defined</u> the characteristics of the o-e transition, there have been several references to the "extraordinary" properties of other polyion systems in which a "splitting" of values in D_{app} occurred as a function of added salt or polyion concentration if salt was not added to the system. The "anomalously slow" relaxation mode has been referred to as the "extraordinary behavior" of the system, and reference was made to the work on PLL and PSS. Not all of the D_{app} versus C_s profiles, however, exhibit the characteristics of the o-e transition as originally defined for the PLL system. It has been suggested that the profiles can be separated into three distinct classes *(79)*. Both Class 1 and Class 2 profiles are defined by a "gradual" splitting, i.e., over several decades of C_s, of D_{app} from a common origin as C_s is lowered into D_{fast} (increasing in value) and D_{slow} (decreasing in value). Class 1 profiles are defined by the constraint that $(D_{fast}/D_{slow})_{max} < 5$, whereas the ratio D_{fast}/D_{slow} is unconstrained for Class 2 profiles. The separation of these two classes was based on the possible interpretation of the data. Class 1 profiles can be explained as a slight polydispersity in the molecular weight of the sample, where the behavior of D_{fast} and D_{slow} may reflect the relative responses of the flexible polyions to chain expansion and osmotic susceptibility changes. Class 2 profiles may result from <u>strongly</u> coupled modes of the macroions of the polydisperse system, where the magnitude of the splitting reflects the off-diagonal coupling constant in the diffusion tensor. Class 3 profiles exhibit a <u>sudden</u> appearance of D_{slow} as C_s is decreased, with $D_{fast}/D_{slow} > 10$. The primary example of the Class 3 profile is the o-e transition in

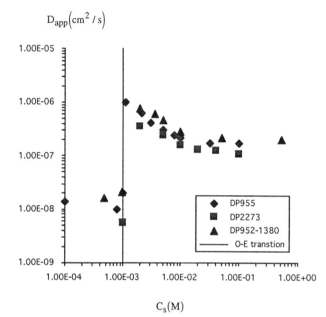

Figure 1. The o-e transition in Poly(L-lysine). The apparent diffusion coefficient D_{app} is plotted as a function of the electrolyte concentration C_S for poly(L-lysine) with various degrees of polymerization (DP). The location of the o-e transition is indicated by a vertical line. [▲ from Figure 1 of Ghosh, Peitzsch, and Reed (*75*); ◆ from Table I of Lin, Lee, and Schurr (*70*); ■ from Table I of Wilcoxon and Schurr (*71*).]

the PLL system. In the Class 3 profiles D_{fast} may be described as a continuous function of C_S whereas D_{slow} exhibits an apparent discontinuity at a well-defined value of C_S that is dependent upon C_m. The D_{app} versus C_S profiles for copolymers of acrylamide/sodium acrylate (PAmA) (*80*) (Class 1), proteoglycan monomers (PG) (*81*) (Class 2), and quaternized poly(2-vinylpyridine) (P2VP) (*82*) (Class 3) are given in Figure 2.

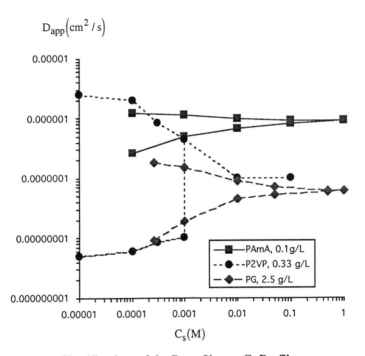

Figure 2. Three Classifications of the D_{app} Versus C_S Profiles.
■   Class 1, represented by poly(acrylamide/acrylate) data from Figure 2 of Reed, Ghosh, Medjahdi, and Francois (*80*).
◆   Class 2, represented by proteoglycan monomer data from Figure 9 of Li and Reed (*81*).
●   Class 3, represented by quaterized poly(2-vinylpyridine) data from Figure 15 of Förster, Schmidt, and Antonietti (*82*).

Interpretation of the Class 3 Profiles. Both Class 1 and Class 2 profiles exhibit a continuous change in both D_{fast} and D_{slow} as C_S is lowered. Hence these profiles can be interpreted by a clever researcher in terms of continuous functions related to the osmotic susceptibility, relative scattering powers of the constituent macromolecules, and/or differential expansion properties of the polyions. What is difficult to comprehend within the context of the current polyelectrolyte paradigm is both the <u>sharpness</u> and the <u>magnitude</u> of the D_{app} versus C_S profile exhibited by Class 3 systems, in particular the o-e transition of PLL and PSS. Indeed the explanation of the o-e transition is currently a controversial topic in the literature, including some of the papers in this symposium-based volume. It is nether the intent to review the many studies in the literature on the o-e transition, nor to examine in detail the attributes any one interpretation of this phenomenon. Attention is thus directed to the salient features of three models of the o-e transition that have been advanced in the literature: "filterable aggregates", "temporal aggregates", and "cluster domains."

The Filterable Aggregate Interpretation. Ghosh, Peitzsch, and Reed (*75*) summarized most of their work on several polyelectrolyte systems and described in detail a "filterable aggregate" (FA) interpretation to explain <u>all</u> of the anomalous slow modes in Classes 1-3 described above. The basic assumption is that a relatively small population of filterable aggregates exist in most polymer preparations. As the ionic strength is lowered the increase in the osmotic susceptibility suppresses only the light scattered by the "monomer" units, thus revealing the aggregated material to the point that it dominates in the low ionic strength region. It has previously been pointed out that the FA model is inconsistent with some of the experimental observations related to the o-e transition of PLL (*83*). For example, Bruno and Mattice (*84*) reported Förster energy transfer studies on PLL with the intent to determine the optimal distance between the donor-acceptor pair. The o-e transition was identified by a sharp increase in the intensity ratio I_{585}/I_{515} at [NaCl] = 10^{-3} M. It was concluded from this study that the optimal distance was in the range 60-100 Å, which was noted to be comparable to the average interparticle spacing deduced from X-ray scattering data.

Even though data exist that are inconsistent with the FA model, it is nonetheless important to examine the virtues of this model in regard to the o-e transition of PLL. The colorful metaphor for this model is that of "stars" (the aggregates, denoted by EP for "extraordinary phase") and the "sun" (the monomer particles, denoted by OP for "ordinary phase"). These authors thus envision the o-e transition as follows (*75*),

> The weak EP scattering from the aggregates (the "stars") is detectable against the dark (extremely weakly scattering) background of the ordinary polyelectrolyte populations at low ionic strength. As C_a increases, the scattering from the ordinary polyelectrolytes rises dramatically (the sun comes up and the atmosphere scatters brightly), and overwhelms the feeble scattering from the aggregates of the EP phase.

Reed and co-workers state that the aggregated material can always be filtered from the solution if one uses a membrane of sufficiently small pore size.

In the context of the "star-sun" metaphor D_{app} is an amplitude weighted value,

$$D_{app} = \frac{A_{EP}D_{EP} + A_{OP}D_{OP}}{A_{EP} + A_{OP}} \tag{29}$$

where the scattering amplitude A_{OP} is suppressed as the osmotic susceptibility is increased. To be consistent with the *apparent discontinuity* in the profiles in Figure

1, it follows that $A_{EP} \ll A_{OP}$ for $[NaBr] > 1.1 \times 10^{-3}$ M and $A_{EP} \gg A_{OP}$ for $[NaBr] < 9.1 \times 10^{-4}$ M. The sharpness of the variation in the relative amplitudes of the two components is an experimental fact that must be explained by any model for the o-e transition. It is thus necessary to verify the relative magnitudes of the change for the FA model before pursuing the task of explaining the sharpness of the transition.

The relative magnitudes of the scattered light intensity attributed to the OP and EP particles in the extraordinary regime can be estimated from the data of Ghosh et al (75) on PLL. The measured intensities of the sample filtered through 0.22 μm membrane filters (EP and OP particles are present) at added salt concentrations 0 and C_{add} are defined as $I_{tils}(0)$ and $I_{tils}(C_{add})$. According to these authors filtration through the 0.05 μm. membrane filters removes the EP particles. Hence $I_{OP}(0)$ is the measured intensity of the EP-free solution at $C_{add} = 0$. They then represented their measurements in terms of a "normalized" intensity, $I_e'(C_{add})$,

$$I_e'(C_{add}) = \frac{I_{tils}(C_{add}) - I_{OP}(0)}{I_{tils}(0) - I_{OP}(0)} \tag{30}$$

The relative intensity of light scattered by the EP particles to the OP particles in zero added electrolyte solvent, $I_{EP}(0)$ and $I_{OP}(0)$, respectively, can be calculated from equation 30 using the Ghosh et al. (75) measured values $I_{tils}(\text{high salt}) = 8 I_{tils}(0)$ and calculated values $I_e'(\text{high salt}) = 40 I_e'(0) = 40$. Such a calculation gives (79),

$$I_{EP}(0) = \frac{7}{32} I_{OP}(0) \tag{31}$$

In other words, the "sun" scatters approximately 5 times more light than the "stars" in the zero added electrolyte solution where the filterable aggregate model requires that only the "stars" be visible.

Aggregation is a potential problem in the interpretation of light scattering data, as emphasized by Reed and co-workers. The discussion above does not diminish the importance of taking steps to remove aggregated material, but rather that the filterable aggregates model is not consistent with the experimental data on the o-e transition.

"Temporal Aggregate" and "Cluster Domain" Interpretations. The descriptions of the "temporal aggregate" (TA) (85) and "cluster domain" (86) models are quite similar, hence it may not be possible to experimentally distinguish between them. [The original phrase used by Sedlák et al. (86) is "interchain domains (clusters)" which was altered to be "cluster domain" (79).]. Both the TA and CD models interpret the fast mode in terms of small ion-polyion coupled mode theory. It is noted, however, that Förster et al. (82) interpret the fast mode as a "gel mode" due to intertwining chains.

The slow mode in the TA model is envisioned to reflect the dynamics of polyions that are coupled due to the "sharing" of the common electrolyte ions in their overlapping ion clouds. The TA model reinterprets the Drifford/Dalbiez ratio (cf. equation 27) in terms of the volume ratio of the Debye-Hückel cloud per unit charge, V_{DH}, to the volume of the monomer unit, V_m, i.e. (87),

$$\rho_{DD} = 4 \frac{\pi \langle b \rangle \lambda_{DH}^2}{Z_s} \frac{C_m^{o-e} N_A}{1000} = 4 \frac{V_{DH}}{V_m} \tag{32}$$

The nature of the dynamics of the system for $\rho_{DD} > 1$ is statistically different than solutions for which $\rho_{DD} < 1$ because of the more active role of the electrolyte ions. The temporal aggregates are therefore *metastable*. The anomalously slow relaxation time may reflect translation of the TA, the kinetics of dissolution of the temporal structure, or a combination of both.

The CD model envisions the slow mode as the translational diffusion of an "interchain domain" whose dimensions can be obtained from the angle dependence of either static or dynamic light scattering (*86*). These domains are formed in the semidilute solution regime, i. e., when the polyion chains physically overlap. These "clusters" of polyions, however, are not "hard aggregates" and the dimensions can be altered, for example, by filtration (*88*).

The importance of the polyion charge in establishing the slow mode is brought out by recent experiments by Sedlák (*88*) on partially ionized poly(methacrylic acid) (PMA) in salt free solutions. In these experiments the neutral PAM ($\alpha=0$) was first filtered through 0.1μm Nucleopore filter. CONTIN analysis of the correlation function showed a large amplitude "fast" peak with a small amplitude "slow" peak. The sample was then titrated while in the scattering cell by likewise filtering through 0.1μm Nucleopore filters appropriate aliquots of a standard NaOH solution to $\alpha=0.4$. The CONTIN analysis now showed a tremendous increase in the relative amplitude of the "slow" peak. The charged PAM was then removed from the scattered cell and re-filtered through a 0.1μm membrane back into the scattering cell. The CONTIN profile for the re filtered sample was similar to that of the previous run. Hence the presence of the slow mode is strongly dependent upon the charged state of the polyion and is not a permanent "filterable aggregate" caused by entanglements.

Comments on the o-e transition. It is not intuitively obvious that the "splitting" phenomenon observed for polyions in "salt-free" solutions is the same as the o-e transition reported for polyions at a fixed concentration as the electrolyte concentration is varied. Förster *et al.* (*82*) reported both types of studies on P2VP. The apparent discontinuity in the D_{app} versus C_S profile was present (cf. Figure 2) whereas the D_{app} versus C_1 profile exhibited a "continuous" splitting reminiscent of the Class 2 systems. In order to better understand the relationship between these two profiles it seems imperative that both kinds of studies be carried out simultaneously on the same sample. It is likewise important to take into consideration the overlap concentration of the polyions, in going from dilute to semidilute solution conditions.

Concluding Comments

We have briefly examined only six topics in the field of polyelectrolytes that have been applied with varying degrees of consistency in the literature. To some, the chosen examples may not be as interesting as the more complex systems such as mixed polyion solutions, adsorption on polyions onto heterogeneous surfaces, the effect of the gel stiffness and charge on the movement of a macroion through the matrix in the presence of an electric field, or the rate of chemical reactions involving polyions, to name only a few. The interpretation of data for these systems also contain some aspects of ambiguity. Nonetheless progress towards the understanding of the fundamental nature of these complex systems has been made, even though precise quantitative agreement with theory may not be achieved. If I may paraphrase a memorable comment made by Dirk Stigter during his talk at this conference,

> There was such good agreement between theory and experiment that we looked for compensating errors.

References

1. Langmuir, I. *J. Chem. Phys.* **1938,** *6*, 873-896.
2. Verwey, E. J. W.; Overbeek, J. Th. G., *Theory of the Stability of Lyophobic Colloids*, Elsevier: Amsterdam, 1948, 135-185.
3. Ise, N.; Okubo, T., *J. Phys. Chem.*, **1966,** *70*, 2400-2405.
4. Sugimura, M.; Okubo, T.; Ise, N.; Yokoyama, S., *J. Am. Chem. Soc.*, **1984,** *106*, 5069-5072.
5. Ise, N.; Okubo, T.; Ito, K.; Dosho, S.; Sogami, I., *Langmuir*, **1985,** *1*, 176-177.
6. Matsuoka, H.; Murai, H.; Ise, N., *Phys. Rev. B*, **1988,** *37*, 1368-1375.
7. Ito, K.; Okumura, H.; Yoshida, H.; Ueno, Y.; Ise, N., *Phys. Rev. B*, **1988,** *37*, 10852-10859.
8. Ito, K.; Nakamura, H.; Yoshida, H.; Ise, N., *J. Am. Chem. Soc.*, **1988,** *110*, 6955-6963.
9. Ise, N.; Matsuoka, H.; Ito, K.; Yoshida, H., *Faraday Discuss. Chem. Soc.*, **1990,** *90*, 153-162.
10. Ise, N.; Matsuoka, H.; Ito, K.; Yoshida, H.; Yamanaka, J., *Langmuir*, **1990,** *6*, 296-302.
11. Sogami, I.; Ise, N. *J. Chem. Phys.* **1984,** *81*, 6320-6332.
12. Overbeek, J. Th. G., *J. Chem. Phys.*, **1987,** *87*, 4406-4408.
13. Smalley, M. *Molec. Phys.*, **1990,** *71*, 1251-1267.
14. Medin a-Noyola, M.; McQuarrie, D. A. *J. Chem. Phys.*, **1980,** *73*, 6279-6283.
15. Lin, S.-C.; Lee, W. I.; Schurr, J. M. *Biopolymers* **1978,** *17*, 1041-1064.
16. Beresford-Smith, B.; Chan, D. Y. C.; Mitchell, D. J. *J. Colloid and Interf. Sci.* **1985** *105*, 216-234.
17. Belloni, L. *J. Chem. Phys.* **1986,** *85*, 519-526.
18. Alexandrowicz, Z.; Katchalsky, A. *J. Polym. Sci. Part A*, **1963,** *1*, 3231-3260.
19. Russel, W. B. *J. Polym. Sci.* **1982** *29*, 1233-1247.
20. Sogami, I. In *Ordering and Organisation in Ionic Solutions*, Editors, Ise, N.; Sogami, I., Eds.; World Scientific Publ.: Teaneck, N. J., 624-634.
21. Schmitz, K. S. *Polym.* **1990,** *31*, 1823-1830.
22. Fowler, R. H. *Trans. Faraday Soc.* **1927,** *23*, 434-443.
23. Oosawa, F.; Imai, N. *J. Chem. Phys.* **1954,** *22*, 2084-2085.
24. Oosawa, F.; Imai, N.; Kagawa, I. *J. Polym. Sci.* **1954,** *13*, 93-111.
25. Oosawa, F. *J. Polym. Sci.* **1957,** *23*, 421-430.
26. Ohnishi, T.; Imai; N.; Oosawa, F. *J. Phys. Soc. Jpn.* **1960,** *15*, 896-905.
27. Oosawa, F. *Polyelectrolytes*, Marcel Dekker, New York, New York, 1971.
28. Manning, G. S. *J. Chem. Phys.* **1965,** *43*, 4260-4267.
29. Manning, G. S. *J. Chem. Phys.* **1969,** *51*, 924-933.
30. Manning, G. S. *J. Chem. Phys.* **1969,** *51*, 934-938.
31. Manning, G. S. *Biophys. Chem.* **1977,** *7*, 95-102.
32. Manning, G. S. *Biophys. Chem.* **1978,** *8*, 65-70.
33. Manning, G. S. *Quart. Rev. Biophys.* **1978,** *11*, 179-246.
34. Manning, G. S. *Acc. Chem. Res.* **1979,** *12*, 443-449.
35. MacGillivray, A. D. *J. Chem. Phys.* **1972,** *56*, 80-82.
36. MacGillivray, A. D. *J. Chem. Phys.* **1972,** *56*, 83-85.
37. Lampert, M. A.; Crandall, R. S. *Chem. Phys. Lett.* **1980,** *72*, 481-485.
38. Skolnick, J.; Grimmelmann, E. K. *Macromolecules* **1980,** *15*, 335-338.
39. Woodbury, Jr., C. P.; Ramanathn, G. V. *Macromolecules* **1982,** *15*, 82-86.
40. Skolnick, J.; Fixman, M. *Macromolecules* **1978,** *11*. 867-871.
41. Dewey, T. G. *Biopolymers* **1990,** *29*, 1793-1799.
42. Anderson, C. F.; Record, Jr., M. T.; Hart, P. A. *Biophys. Chem.* **1978,** *7*, 301-316.
43. Bleam, M. L.; Anderson, C. F.; Record, M. T., Jr. *Proc. Natl. Acad. Sci. USA* **1980,** *77*, 3085-3089.

44. Rose, D. M.; Bleam, M. L.; Record, M. T., Jr.; Bryant, R. G. *Proc. Natl. Acad. Sci. USA* **1980**, *77*, 6289-6292.
45. Granot, J.; Kearns, D. R. *Biopolymers* **1982**, *21*, 219-232.
46. Braunlin, W. H.; Anderson, C. F.; Record, M. T., Jr. *Biochemistry*, **1987**, *26*, 7724-7731.
47. Braunlin, W. H.; Xu, Q. *Biopolymers*, **1992**, *32*, 1703-1711.
48. Rhee, K. W.; Ware, B. R. *J. Chem. Phys.* **1983**, *78*, 3349-3353.
49. Klein, J. W.; Ware, B. R. *J. Chem. Phys.* **1984**, *80*, 1334-1339.
50. Le Bret, M.; Zimm, B. H. *Biopolymers* **1984**, *23*, 287-312.
51. Schmitz, K. S. *Macroions in Solution and Colloidal Suspension* VCH Publishers: New York, NY, **1993**, 243-251.
52. Matsumoto, M.; Kataoka, Y. In *Ordering and Organisation in Ionic Solutions*, Editors, Ise, N.; Sogami, I., Eds.; World Scientific Publ.: Teaneck, N. J., 574-582.
53. Kirkwood, J. G.; Schumaker, J. B. *Proc. Natl. Acads. Sci. USA* **1952**, *38*, 863-871.
54. Oosawa, F. *Biopolymers* **1968** 6, 1633-1647.
55. Fulton, R. L. *J. Chem. Phys.* **1978**, *68*, 3089-3094.
56. Fulton, R. L. *J. Chem. Phys.* **1978**, *68*, 3095-3098.
57. Schmitz, K. S.; Parthasarathy, N.; Vottler, E. *Chem. Phys.* **1982**, *66*, 187-196.
58. Schmitz, K. S. ; Parthasarathy, N.; Kent, J. C.; Gauntt, J *Biopolymers* **1982**, *21*, 1365-1382.
59. Schmitz, K. S.; Lu, M.; Gauntt, J. *J. Chem. Phys.* **1983**, *78*, 5059-5066.
60. Yoshino, S. In *Ordering and Organisation in Ionic Solutions*, Editors, Ise, N.; Sogami, I., Eds.; World Scientific Publ.: Teaneck, N. J., 449-459.
61. Manning, G. S. Invited lecture at the 205th National Am. Chem. Soc. Meeting, Denver, CO, 1993.
62. Sánchez-Sánchez, J. E.; Lozada-Cassou, M. *Chem. Phys. Letts.* **1992**, *190*, 202-208.
63. Odijk, T. *J. Polym. Sci. Polym. Phys. Ed.* **1977** 15, 477-483.
64. Skolnick, J.; Fixman, M. *Macromolecules*, **1977** 10, 944-948.
65. Odijk, T.; Houwaart, A. C. *J. Polym. Sci. Polym. Phys. Ed.* **1978** 16, 627-639.
66. Le Bret, M *J. Chem. Phys.* **1982** 76, 6243-6255.
67. Fixman, M. *J. Chem. Phys.* **1982** 76, 6346-6353.
68. Tricot, M. *Macromolecules* **1984** 17, 1698-1704.
69. Barrat, J.-L.; Joanny, J.-F. (submitted to *Europhysics Lett.*).
70. Lin, S.-C.; Lee, W. I.; Schurr, J. M. *Biopolymers* **1978**, *17*, 1041-1064.
71. Wilcoxon, J. P.; Schurr, J. M. *J. Chem. Phys.* **1983**, *78*, 3354-3364.
72. Zero, K.; Ware, B. R. *J. Chem. Phys.* **1984**, *80*, 1610-1616.
73. Schmitz, K. S.; Lu, M.; Singh, N.; Ramsay, D. J. *Biopolymers* **1985**, *23*, 1637-1646.
74. Ramsay, D. J.; Schmitz, K. S. *Macromolecules* **1985** ,18, 2422-2429.
75. Ghosh, S.; Peitzsch, R. M.; Reed, W. F. *Biopolymers* **1992**, *32*, 1105-1122.
76. Drifford, M.; Dalbiez, J.-P. *Biopolymers* **1985**, *24*, 1501-1514.
77 Martin, N. B.; Tripp, J. B.; Shibata, J. H.; Schurr, J. M. *Biopolymers* **1979**, *18*, 2127-2133
78 Shibata, J. B.; Schurr, J. M. *Biopolymers* **1979**, *18*, 1831-1833.
79. Schmitz, K. S. *Biopolymers* **1993** 33 , 953-959.
80. Reed, W. F.; Ghosh, S.; Medjahdi, G.; Francois, J. *Macromolecules* **1991**, *24*, 6189-6198.
81. Li, X.; Reed, W. F. *J. Chem. Phys.* **1991**, *94*, 4568-4580.
82. Förster, S.; Schmidt, M.; Antonietti, M. *Polymer* **1990**, *31*, 781-792.
83. Schmitz, K. S. *Macroions in Solution and Colloidal Suspension* VCH Publishers: New York, NY, **1993**, 340-348.
84. Bruno, K.; Mattice, W. L. *Macromolecules* **1992**, *30*, 310-330.

85. Schmitz, K. S.; Lu, M.; Singh, N.; Ramsay, D. J. *Biopolymers* **1984** *23*, 1637-1646.
86. Sedlák, M.; Konák, C.; Stepánek, P.; Jakes, J. *Polymer*, **1987**, *28*, 873-880.
87. Schmitz, K. S.; Ramsay. D. J. *J. Colloid and Interf. Sci.* **1985**, *105*, 388-398.
88. Sedlák, M. *Macromolecules* **1993**, *26*, 1158-1162.

RECEIVED August 6, 1993

THEORY

Chapter 2

Ion–Ion Correlations in the Electrical Double Layer around a Cylindrical Polyion

J. Reščič and V. Vlachy[1]

Department of Chemistry, University of Ljubljana, 61000 Ljubljana, Slovenia

Ion-ion correlations in the electrical double-layer around a cylindrical polyion are studied using the grand canonical Monte Carlo technique. The results for the ion-ion distribution functions are used to examine the effect of DNA on bimolecular chemical reaction rates. The comparison with experimental data for the energy transfer in DNA-electrolyte mixtures indicates, that the theory significantly overestimates the effects of DNA on the counterion-counterion collision frequency.

The clustering of counterions around linear polyelectrolytes in the solution has long been theoretically predicted (*1-3*). This phenomenon has important consequences for thermodynamic and transport properties of polyelectrolyte solutions. Theoretical studies, published so far, are primarily concerned with the distribution of ions around polyion (*1*) (singlet distribution function) and little attention is paid to the ion-ion distributions (two-particle distribution functions) in the cylindrical double-layer. Two-particle distribution functions are rarely studied even for a simpler, planar, geometry (*4*). The study of ion-ion correlations in the cylindrical double-layer presented here, is motivated by the experimental work of Wensel and co-workers (*5*), who measured the effect of DNA on rates of bimolecular energy transfer between small ions. These measurements provide a direct indication of how the ions are distributed in the electrical double-layer around the polyion. The measured effect of DNA on ion-ion collision frequencies has been used to test various polyelectrolyte theories, like counterion-condensation theory (*2*) and the Poisson-Boltzmann cell theory (*1,5*). The conclusion is that both, in general quite successful polyelectrolyte theories, grossly overestimate the effect of DNA on rates of energy transfer. The analysis of experimental data presented in Reference (*5*) is based on the equation of Morawetz (*6*), where the observed rate constants for energy transfer in the presence

---

[1]Corresponding author

0097–6156/94/0548–0024$06.00/0

of a polyelectrolyte k and its absence, k_0, are related as:

$$\frac{k}{k_0} = \frac{V^{-1} \int n_i(r) \, n_j(r) dV}{[\, V^{-1} \int n_i(r) dV \,] \, [\, V^{-1} \int n_j(r) dV \,]} \tag{1}$$

The local number concentrations of small ions, $n_i(r)$, are evaluated by the Poisson-Boltzmann equation. The domain of integration extends over all volume accessible to small ions. The shortcomings of the Poisson-Boltzmann theory have been analyzed in many papers, most recently perhaps in References (7-9). In short, the Poisson-Boltzmann approximation treats the small ions as pointlike charges, ignoring their mutual correlations (10). It should be noted that equation 1 contains similar mean-field type of approximations, as inherent to the Poisson-Boltzmann theory.

In the present paper we suggest an alternative approach to analyze the effect of polyelectrolyte upon reaction rates. The grand canonical Monte Carlo method is used to study the correlations between small ions in the double-layer around the DNA polyion. If DNA is added to an electrolyte solution, the counterions concentrate in the vicinity of DNA molecules. The effect of DNA on ion-ion correlations is assumed to be a measure of the increase of bimolecular energy transfer rate in the DNA-electrolyte mixtures (5). During the Monte Carlo simulation, the spherically symmetric ion-ion correlation functions $g_{ij}(r)$, not studied in previous calculations, are monitored. The ratio k / k_0 is approximated by (11)

$$\frac{k}{k_0} = \frac{\langle g_{AB}(a) \rangle}{g_{AB}^0(a)} \tag{2}$$

where a is the distance of collision of the two ions and $\langle ... \rangle$ denotes the volume average. The Monte Carlo results for equation 2 are compared with the results of the Morawetz equation (with local concentrations calculated from the Poisson-Boltzmann theory) and with experimental data of Reference (5).

Models and Methods

Cell Model. The model of DNA-electrolyte mixture used in this study treats the polyion as an impenetrable, infinitely long cylinder. Each polyion is placed along the long axis of a cylindrical cell. A collection of such, identical and independent cells, is taken to represent as solution. The radius R of the cell is related to the concentration of the solution C_m, expressed in moles of the monomer units.

$$C_m = \frac{1}{\pi \, (R^2 - \sigma^2) \, b \, N_A} \tag{3}$$

where b is the length of the monomer unit, σ is the distance of closest approach between ion and polyion and N_A is Avogadro's number. The fixed charge, one per length b, is assumed to be smeared uniformly over the polyion surface. The counterions and co-ions, in Monte Carlo calculation they are modeled as charged hard spheres of diameter a, move in a dielectric continuum. Moreover, the relative permittivity ϵ_r is assumed to be uniform through the system.

Poisson-Boltzmann Equation. The Poisson-Boltzmann equation for the cylindrical symmetry reads

$$\frac{1}{r}\frac{d}{dr}(r\frac{d\psi(r)}{dr}) = -\frac{e_0}{\varepsilon_0\varepsilon_r}\sum_i z_i n_i(r)$$

(4)

$$n_i(r) = n_i(R)exp[-z_i e_0 \psi(r)]$$

where $\psi(r)$ is the mean electrostatic potential at a distance r and $n_i(0)$ is the number concentration of ionic species i at $r = R$. As usual $\beta = (k_B T)^{-1}$, where k_B is the Boltzmann's constant and T absolute temperature. The appropriate boundary conditions, given by the Gauss Law, are

$$(\frac{d\psi(r)}{dr})_R = 0 \; , \; (\frac{d\psi(r)}{dr})_\sigma = \frac{e_0}{2\pi \; \varepsilon_0 \varepsilon_r \; \sigma \; b}$$

(5)

The Poisson-Boltzmann equation has been solved subject to the boundary conditions using the so-called "shooting method" (12). Once the mean electrostatic potential is known, the volume averages needed in equation 1 can readily be calculated.

Grand Canonical Monte Carlo Calculation. The grand canonical Monte Carlo method has been introduced in the electrical double-layer studies by Valleau and co-workers (13,14). The mixture of electrolyte and DNA is assumed to be separated from "bath" of pure electrolyte (bulk) by a semipermeable membrane. In equilibrium, the distribution of small ions (polyions cannot cross the membrane) between DNA-electrolyte mixture and bulk electrolyte solution is given by equality of the chemical potential on each side. The advantage of the method is that by sampling at constant chemical potential, the relevant bulk phase is defined unambiguously. The method has been used and described in detail in several papers (9,13-16).

The simulation procedure consists of two steps. The first step is canonical: a randomly chosen ion is moved into a new random position in the Monte Carlo cell. The attempted move is accepted with probability f_{ij} (equation 6), where U_i (U_j) is the configurational energy of state i (j), and $Y = 1$. In the next step, a random decision is made to either attempt the insertion or deletion of a neutral combination of ions. The transition probability from the state i (N^-_i, N^+_i) to state j (N^-_j, N^+_j) is given by combination of equations 6 and 7:

$$f_{ij} = min \{1 , Y \ exp[-\beta \ (U_j - U_i) \ ] \}$$ (6)

$$Y = \frac{N^+ \ N^- \ \gamma_{\pm,s}}{N_j^+ \ N_j^-} \ \textit{(addition)} \ ; \ Y = \frac{N_i^+ \ N_i^-}{N^+ \ N^- \ \gamma_{\pm,s}} \ \textit{(deletion)}$$ (7)

γ_{+-} is the mean activity coefficient of bulk electrolyte with ionic concentrations N_+/V and N_-/V. The corrections due to the finite size of the Monte Carlo cell, which may be important for anisotropic systems, have been evaluated using the procedure developed by Rossky and co-workers (*17*). In this calculation the number of monomer units b included in the Monte Carlo cell is from 90 to 180 depending on the polyelectrolyte concentration.

Results

In this section we present numerical results for the model system described above. Counterions and co-ions are of equal size, diameter a = 0.8 nm and σ = 1.38 nm. All calculations apply to water-like solutions at T = 298 K. The central parameter of the Poisson-Boltzmann cell theory, $\lambda = \beta e^2/(4\pi\epsilon_r\epsilon_0 b)$ = 4.2. The grand canonical simulation requires as input the mean activity coefficient of the pure (bulk) electrolyte solution. The results for $\gamma_\pm$ at various concentrations are obtained separately and they are given in Table I.

Table I. Numerical Results for the Model System

$C_s 10^3/(mol/dm^3)$	1.24	1.99	2.47	3.60	5.00
$\gamma_\pm$	0.966	0.957	0.955	0.949	0.943
$C_s 10^3/(mol/dm^3)$	6.32	10.18	28.05	49.92	100.1
$\gamma_\pm$	0.940	0.931	0.935	0.960	1.051

The two-particle distribution function $g_{ij}(r_i,r_j) = g_{ij}(r_i,r_{ij})$ depends on the positions r_i and $r_j = r_i + r_{ij}$, of the two ions in the cylindrical cell. We are interested in the correlation function $g_{ij}(r_{ij}) = g_{ij}(|r_i - r_j|)$, which measures the probability for two ions to be within the distance r, r + dr from each other. During the computer simulation, $g_{ij}(r_i,r_{ij})$ is averaged over the position r_i to obtain $g_{ij}(r_{ij})$. In order to accumulate good statistics very long Monte Carlo runs are needed. The distribution functions are collected on the grid of 0.1 nm and averaged over 3000 or more configurations per particle.

In Figure 1 we present the correlation function $g_{++}(r)$ for counterions in the DNA-electrolyte mixture with C_m = 0.01 monomol/dm³ and C_s = 0.002 mol/dm³. Small enhancement in $g_{++}(r)$ is due to the clustering of counterions around the DNA molecule. The same effect has been observed in solutions of highly asymmetric electrolytes (18-20). The equivalent correlation function for bulk electrolyte of concentration C_s is shown in the same figure. The addition of polyelectrolyte dramatically changes the counterion-counterion correlations and it is expected to have a strong effect on the chemical reaction rate. The similar holds true for the counterion-coion pair (+,-) as presented in Figure 2, while the equivalent function for co-ions (not shown here), $g_-(r)$, is almost unaffected by the presence of polyelectrolyte. The results for electrolyte-polyelectrolyte mixture with $C_m = 0.002$ monomol/dm^3 and $C_s = 0.002$ mol/dm^3 are shown in Figures 3 and 4. Finally, the results for the case where simple electrolyte is added in an excess, i.e. $C_m = 0.0004$ monomol/dm^3 and $C_s = 0.002$ mol/dm^3, are presented Figures 5 and 6. In this case the correlation function for +,- pair, $g_\pm(r)$, shows almost no change upon addition of polyelectrolyte in the system.

The theory of chemical reactions suggests that reaction rates may be related to the static structure of the solution. Under certain approximations the rate for bimolecular reaction is proportional to $g_{ij}(a)$, where a is a distance of closest approach of two ions (11). The ratio of the energy-transfer rates between counterions, calculated by equation 2 at several DNA concentrations, is shown in Figure 7. In the same figure we show the results obtained from equation 1 and the solution of the Poisson-Boltzmann equation (line) and the experimental data compiled from Figure 3 of Reference (5). It is quite clear from this comparison that equation 2 (and Monte Carlo simulation) yields a slightly better agreement with the experiment, but the agreement is far from being quantitative. Both theories, based either on equation 1 or 2, greatly overestimate the effects of DNA on collision frequencies at low DNA concentrations. For example, at $C_m = 0.0004$ mole/dm^3 and $C_s = 0.002$ mole/dm^3 equation 1 (Poisson-Boltzmann theory) and equation 2 (Monte Carlo simulation) give $k / k_0 = 35$ and 22, respectively, while the experimentally obtained value for these concentrations is much lower, around 3. Finally we need to mention, that evaluation of $g_{ij}(a)$ requires an extrapolation of the correlation function, what is associated with a measurable uncertainty, as shown in Figures 1-6.

Conclusions

In 1988 Wensel and coworkers (5) presented an interesting experimental study of DNA-electrolyte mixtures. Initial theoretical analysis, based on the Poisson-Boltzmann theory (5) is only qualitatively correct; the effects of DNA on the rate of energy transfer between counterions is much smaller than theoretically predicted. The similar conclusion seems to apply to the condensation theory of Manning, though the possibilities of this theory have not been fully explored yet. The third calculation used in analysis of the energy transfer data is a semiempirical theory suggested by Matthew (5). This approach, which uses the molecular structure of B-DNA, yields better agreement with measured data than other two theories, but it also contains more parameters.

In this paper we have explored an alternative approach, where the ion-ion correlations between small ions in the cylindrical double-layer are studied by the computer simulation method. No such data are taken in previous Monte Carlo studies of polyelectrolyte solutions. The correlation function between counterions is

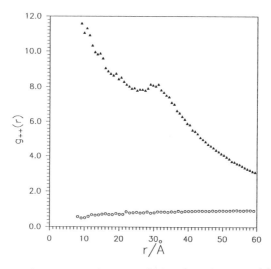

Figure 1. Counterion–counterion correlation function, $g_{++}(r)$, for DNA–electrolyte mixture (▲), and for pure electrolyte solution (○). The concentrations are: $C_m = 0.01$ monomol/dm^3 and $C_s = 0.002$ mol/dm^3.

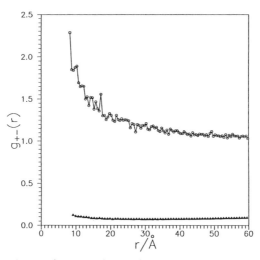

Figure 2. Counterion–coion correlation function, $g+-(r)$, for DNA–electrolyte mixture (▲) and for pure electrolyte solution (○). The concentrations are: $C_m = 0.01$ monomol/dm^3 and $C_s = 0.002$ mol/dm^3.

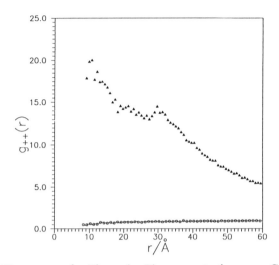

Figure 3. The same as for Figure 1. The concentrations are: $C_m = 0.002$ monomol/dm³ and $C_s = 0.002$ mol/dm³.

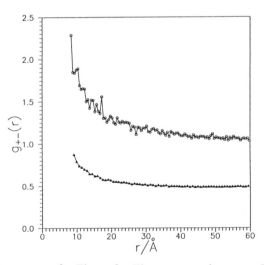

Figure 4. The same as for Figure 2. The concentrations are: $C_m = 0.002$ monomol/dm³ and $C_s = 0.002$ mol/dm³.

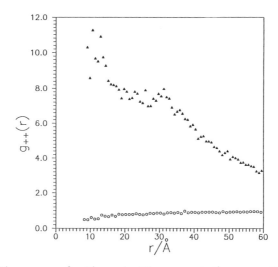

Figure 5. The same as for Figure 1. The concentrations are: C_m = 0.0004 monomol/dm³ and C_s = 0.002 mol/dm³.

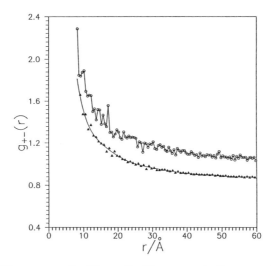

Figure 6. The same as for Figure 2. The concentrations are: C_m = 0.0004 monomol/dm³ and C_s = 0.002 mol/dm³.

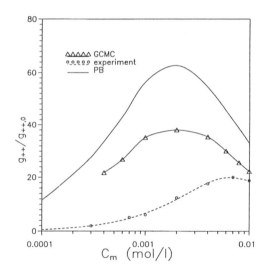

Figure 7. Rate constant ratio k/k_0 for energy transfer between monovalent cations as a function of DNA concentration. Experimental data (broken line) from Reference (5). The uncertainty in the Monte Carlo results is from 5% to 10%.

via equation 2 related to the energy-transfer rate in DNA-electrolyte mixtures. The results are disappointing. The new approach seems to work slightly better than a combination of equation 1 and the Poisson-Boltzmann theory, but there is no quantitative agreement with the experimental data. Particularly poor agreement is obtained for low concentrations of DNA in mixture. The fact that various, very different theories, fail to describe the experimental results seems to suggest that the (rod-like) model used here is a poor representation for the DNA polyion.

Literature Cited

1) Katchalsky, A. *Pure. Appl. Chem.* **1971**, *26,* 327.
2) Manning, G. S. *Acc. Chem. Res.* **1976**, *12,* 443.
3) Anderson, C. T.; Record Jr., M. T. *Annu. Rev. Phys. Chem.* **1982**, *33,* 191.
4) Henderson, D.; Plischke, M. *J. Phys. Chem.* **1988**, *92,* 7177.
5) Wensel, T. G.; Meares, C. F.; Vlachy, V; Matthew, J. B. *Proc. Natl. Acad. Sci. U. S.* **1986**, *83,* 3267.
6) Morawetz, H. *Acc. Chem. Res.* **1970**, *3,* 354.
7) Valleau, J. P.; Ivkov R.; Torrie, G. M. *J. Chem. Phys.* **1991**, *95,* 520.
8) Torrie, G. M. *J. Chem. Phys.* **1992**, *96,* 3772.
9) Jamnik B.; Vlachy, V. *J. Am. Chem. Soc.* **1993**, *115,* 660.
10) Fixman, M.; *J. Chem. Phys.* **1979**, *70,* 4995.
11) Keizer, J. *Acc. Chem. Res.* **1985**, *18,* 235.
12) Carnahan, B.; Luther, H. A.; Wilkes, J. O. *Applied Numerical Methods;* Wiley: New York, U.S., 1969.
13) Torrie, G. M.; Valleau, J. P. *J. Chem. Phys.* **1980**, *73,* 5807.
14) Torrie, G. M.; Valleau, J. P. *J. Phys. Chem.* **1982**, *86,* 3251.
15) Vlachy, V.; Haymet, A. D. J. *J. Chem. Phys.* **1986**, *84,* 5874.
16) Mills, P.; Anderson, C. F.; Record, M. T. *J. Phys. Chem.* **1986**, *90,* 6541.
17) Murthy, C. S.; Bacquet R. J.; Rossky, J. P. *J. Phys. Chem.* **1985**, *89,* 701.
18) Vlachy, V.; Marshall, C.; Haymet, A. D. J. *J. Am. Chem. Soc.* **1989**, *111,* 4160.
19) Reščič, J.; Vlachy, V.; Haymet, A. D. J. *J. Am. Chem. Soc.* **1990**, *112,* 3398.
20) Linse, P. *J. Chem. Phys.* **1990**, *93,* 1376.

RECEIVED August 6, 1993

Chapter 3

A Perturbative Approach to Polyelectrolyte Configuration

D. Bratko[1] and K. A. Dawson[2]

Department of Chemistry, University of California, Berkeley, CA 94720

A simple self-consistent perturbative approach is applied to study the configuration of an isolated ringlike polyion. The calculation is based on the application of Edwards Hamiltonian for a self avoiding macromolecule, supplemented by screened electrostatic interactions among ionized units of the chain. The structure is described in terms of mean square distances between the beads, calculated using the probability distribution of a reference Hamiltonian with purely harmonic interactions. The method provides analytic scaling relations for mean square distances among distant units and lends itself in a form, suitable for numerical solution. The swelling of a strongly ionized polyion is studied as a function of its length L and the ionic strength of the supporting solution. Extension of the method to systems with attractive intramolecular forces between nonadjacent units is considered and numerical results for a helix-forming polymer are presented.

The configuration of an ionized macromolecule in solution is determined by an interplay of chain elasticity, steric effects and long-ranged electrostatic repulsion. These interactions have been the subject of a variety of theoretical (1-21) and simulation (8,22-30) studies of polyelectrolyte solutions. In view of the complexity of real systems, different simplifications were introduced to facilitate the theoretical analysis. A conventional model of polymer theory describes the bonds among monomer units as harmonic springs and treats the excluded volume forces in terms of the Dirac δ function

[1]Also affiliated with J. Stefan Institute, University of Ljubljana, Ljubljana, Slovenia
[2]Current address: Department of Chemistry, University College Dublin, Belfield, Dublin 4, Ireland

(31). Electrostatic effects may be introduced as an additional term in the polymer Hamiltonian, in the simplest case represented by a sum of pairwise additive Yukawa interactions among the charged groups on the chain. In the case of long neutral macromolecules, valuable insights are known to be obtained by studying polymer rings which behave in a similar way as linear chains and, in the case of the above Hamiltonian, lend themselves to a useful analytic solution *(32-34)*. In the preceding works *(21,35)*, a generalization of the ringlike polymer model to charged systems has been suggested and a numerical scheme based on the variational method for the polyelectrolyte structure has been proposed *(21,35,36)*. In the present paper, the variational treatment of these works is shown to be equivalent to a special case of a more general perturbative approach *(37)*. For the particular type of the Hamiltonian, used in our system, the first order perturbation approximation is equivalent to the application of the Gibbs-Bogoliubov bound for the free energy of the polymer. In either approach, the polymer configuration is determined by averaging with respect to an approximate probability distribution of a reference system with a Gaussian Hamiltonian of a collection of Harmonic oscillators. The true Hamiltonian of our system and the perturbative analysis will be described in the following Section. The results will be used to establish scaling relations for mean square distance between distant beads in a number of characteristic situations. In the last Section, we describe the numerical solution of the system of nonlinear equations for reference Hamiltonian couplings and present numerical results for a set of strongly charged dilute polyelectrolytes in a good solvent and at varying degree of polymerization and concentration of the simple electrolyte. Finally, we consider extensions to anisotropic systems and polymers with specific attractive intramolecular interactions.

Model and Method

The cyclic polyelectrolyte molecule is pictured as a necklace comprising L monomer units. The unit length in the absence of steric and electrostatic forces is l. The total charge of the polyion is uniformly spread over all beads, each carrying an average charge q. The bonds between neighbouring units are treated as harmonic springs and we describe the excluded volume interactions in terms of a virial expansion for the Dirac δ function form of the potential between the particles. The electrostatic effects are approximated by the Debye-Huckel screened potential among the beads of the polyion. The solution is sufficiently diluted that there are no significant intermolecular interactions. The configuration dependent part of the Hamiltonian is

$$H = \frac{3}{2l^2} \sum_{s=1}^{L-1} (\mathbf{r}_{s+1} - \mathbf{r}_s)^2 + \frac{u_2}{2} \sum_{s,s'} \delta(\mathbf{r}_s - \mathbf{r}_{s'})$$

$$+ \frac{u_3}{3!} \sum_{s,s',s''} \delta(\mathbf{r}_s - \mathbf{r}_{s'})\delta(\mathbf{r}_s - \mathbf{r}_{s''}) + ... + V\sum_{s,s'} \frac{e^{-\kappa|\mathbf{r}_s - \mathbf{r}_{s'}|}}{|\mathbf{r}_s - \mathbf{r}_{s'}|} \tag{1}$$

where the energy is given in units kT, k is the Boltzmann constant and T the temperature, κ is the Debye screening parameter, $V = q^2/8\pi\varepsilon kT$ is one half of the Bjerrum length, ε the permittivity, s the relative position on the chain and the summations are carried out over all L units of the polyion. At nonzero values of the two particle excluded volume parameter u_2, the three body term can usually be omitted but will become important (35) at theta conditions where it represents the dominant short ranged interaction. The structure of the polyion is described in terms of mean square distances $<|r(s)-r(s')|^2>$ for all separations $|s-s'|$ in the range from 0 to L/2. The averages will be calculated by considering the probability distribution over possible configurations for a reference system with a Gaussian Hamiltonian H_o

$$H_o = \sum_{s<s'} \Gamma(s-s')|r_s - r_{s'}|^2 \qquad (2)$$

where the couplings $\Gamma(s-s')$ have yet to be determined. We proceed to this end by employing a self-consistent perturbative approach, investigated some time ago (37) in a study of a neutral polymer. We begin by writing the canonical average of an observable $O(\{r_s\})$

$$<O>_H = \frac{Tr[e^{-(H-H_o)}e^{-H_o}O]}{Tr[e^{-(H-H_o)}e^{-H_o}]} \qquad (3)$$

Expansion in cumulant series gives

$$<O>_H = <O>_{H_o} - <O(H-H_o)>_{H_o} + <O>_{H_o}<(H-H_o)>_{H_o} + O[(H-H_o)^2] \qquad (4)$$

and we can choose the reference Hamiltonian in such a way that

$$<O(H-H_o)>_{H_o} = <O>_{H_o}<(H-H_o)>_{H_o} \qquad (5)$$

which amounts to the first order perturbation theory, the difference between the averages of the observable O in the true and the reference ensemble being of the order $(H-H_o)^2$. Within this level of approximation, equation 5 can, therefore, be used to determine the couplings $\Gamma(s-s')$ of equation 2. The procedure is simplified ($32,33$) by using the Fourier components (38)

$$\hat{r}(q) = \frac{1}{\sqrt{L}}\sum_s r(s)e^{iqs} \ , \ r(s) = \frac{1}{\sqrt{L}}\sum_q \hat{r}(q)e^{iqs} \qquad (6)$$

with $q=2\pi i/L$, $i=0,1,2...L-1$,

$$\sum_{s<s'} \Gamma(s-s')|\mathbf{r}_s - \mathbf{r}_{s'}|^2 = \sum_q g(q)|\hat{\mathbf{r}}(q)|^2 \tag{7}$$

and

$$\frac{3}{2l^2} \sum_{s=1}^{L-1} (\mathbf{r}_{s+1} - \mathbf{r}_s)^2 = \sum_q j(q)|\hat{\mathbf{r}}(q)|^2 \tag{8}$$

where

$$g(q) = \sum_s \Gamma(s)[1-\cos qs] \quad \text{and} \quad j(q) = \frac{3}{l^2}(1-\cos q) \tag{9}$$

The difference $<H-H_o>$ is obtained by expanding the δ functions in plane waves and then averaging with respect to the reference Hamiltonian

$$<H-H_o> = \sum_q [j(q) - g(q)] <|\hat{\mathbf{r}}(q)|^2> + \frac{\hat{u}_2}{2} \sum_{s,s'} <|\mathbf{r}(s)-\mathbf{r}(s')|^2>^{-\frac{3}{2}} +$$

$$\frac{\hat{u}_3}{3!} \sum_{s,s',s''} \int d\mathbf{k} \int d\mathbf{k}' \, e^{-\frac{k^2}{6}<|\mathbf{r}(s)-\mathbf{r}(s')|^2> - \frac{k'^2}{6}<|\mathbf{r}(s)-\mathbf{r}(s'')|^2>}$$

$$e^{-\frac{\mathbf{k}\mathbf{k}'}{4}[<|\mathbf{r}(s)-\mathbf{r}(s')|^2> + <|\mathbf{r}(s)-\mathbf{r}(s'')|^2> - <|\mathbf{r}(s')-\mathbf{r}(s'')|^2>]}$$

$$\frac{2}{\pi} V \sum_{s,s'} \int_0^\infty \frac{e^{-\frac{k^2}{6}<|\mathbf{r}(s)-\mathbf{r}(s')|^2>}}{k^2+\kappa^2} k^2 dk \tag{10}$$

where $\hat{u}_2 = (3/2\pi)^{\frac{3}{2}} u_2$, and $<|\mathbf{r}(s)-\mathbf{r}(s')|^2>$ is the mean squared distance between the units s and s'.

$$<|\mathbf{r}(s)-\mathbf{r}(s')|^2> = \frac{3}{L} \sum_q \frac{1-\cos q(s-s')}{g(q)} \tag{11}$$

Applying equation 5 to the observables $O=|\hat{\mathbf{r}}(q)|^2$ at $L/2$ discrete values $q_i=2\pi i/L$, i=1,2..L/2, and neglecting the three body or higher terms, we obtain the first order approximation for the unknown coupling coefficients g(q):

$$g(q) - j(q) + \frac{3\hat{u}_2}{2L} \sum_{s,s'} <|r(s) - r(s')|^2>^{-\frac{5}{2}} [1 - \cos q(s - s')] +$$

$$\frac{2V}{3\pi L} \sum_{s,s'} \int_0^\infty \frac{e^{-\frac{k^2}{6} <|r(s) - r(s')|^2>}}{k^2 + \kappa^2} k^4 dk [1 - \cos q(s - s')] + .. . = 0 \qquad (12)$$

Identical result has previously been derived by minimizing the Gibbs-Bogoliubov bound of the free energy of the system (21,35). Applying direct iteration along with the analytic solution to the integral in the electrostatic term, equation 12 can readily be solved at arbitrary conditions (35). In the limits of either very strong or vanishingly weak screening, expansions in terms of k^2/κ^2 can be used to simplify the electrostatic term of equation 12 thus making possible a direct scaling analysis at some extreme conditions. The knowledge of the long distance (large |s-s'|) behavior requires the small q dependence of g(q) which may be shown to vanish as

$$g(q) \underset{q\sim0}{\sim} \frac{1}{D} q^{2\beta} \qquad (13)$$

Equations 11 and 13 give

$$<|r(s) - r(s')|^2> \approx D|s - s'|^{2\alpha} , \qquad 2\alpha = 2\beta - 1 \qquad (14)$$

In view of equations 11-14, equation 12 at large κ simplifies to

$$g(q) - \frac{3}{2l^2} q^2 + \left(a_1 \frac{V}{\kappa^2} + \frac{3}{2} \hat{u}_2 \right) \frac{1}{L} \sum_{s,s'} '[1 - \cos q(s - s')] <|r(s) - r(s')|^2>^{-\frac{5}{2}}$$

$$-\frac{b_1 V}{L\kappa^4} \sum_{s,s'} [1 - \cos q(s - s')] <|r(s) - r(s')|^2>^{-\frac{7}{2}} + ... = 0 \qquad (15)$$

with $a_1 = 9\sqrt{\frac{6}{\pi}}$ and $b_1 = 15a_1$. For small q (q<<κ) and large L, the leading behavior of equation 15 is

$$\frac{1}{D}q^{2\beta} - \frac{3}{2I^2}q^2 + \frac{I_1}{D^{\frac{5}{2}}}\left(a_1\frac{V}{\kappa^2} + \frac{3}{2}\hat{u}_2\right)q^{5\alpha-1} - \frac{b_1 I_2}{D^{\frac{7}{2}}}\frac{V}{\kappa^4}q^{7\alpha-1} + ... = 0 \tag{16}$$

where I_1 and I_2 are pure constants obtained by integration over |s-s'|. Applying the method of dominant balance (39) at small q we find the solution

$$5\alpha - 1 = 2 \ , \ \alpha = \frac{3}{5}, \ D = \left[\frac{2I_1^2}{3}\left(a_1\frac{V}{\kappa^2} + \frac{3}{2}\hat{u}_2\right)\right]^{\frac{2}{5}} \tag{17}$$

Over long distances |s-s'|>>1 a well screened polymer resembles a self-avoiding ring with a modified second virial coefficient. A more careful analysis (35) avoiding the approximations of the preceding paragraph yields similar results but with Reiss exponent $2\alpha=4/3$ replacing the Flory result. At shorter distances or in the absence of simple electrolyte, when we may ignore the screening of Coulombic interactions, an alternative expansion of equation 12 gives

$$g(q) - \frac{3}{2I^2}q^2 + \frac{3\hat{u}_2}{2L}\sum_{s,s'} <|\mathbf{r}(s) - \mathbf{r}(s')|^2>^{-\frac{5}{2}}[1 - \cos q(s-s')] +$$

$$\frac{a_2 V}{L}\sum_{s,s'} '<|\mathbf{r}(s) - \mathbf{r}(s')|^2>^{-\frac{3}{2}}[1 - \cos q(s-s')] +$$

$$-\frac{b_2 V\kappa^2}{L}\sum_{s,s'} '<|\mathbf{r}(s) - \mathbf{r}(s')|^2>^{-\frac{1}{2}}[1 - \cos q(s-s')] + ... = 0 \tag{18}$$

with $a_2 = \sqrt{\frac{6}{\pi}}$ and $b_2 = \frac{a_2}{3}$ The leadind terms of equation 18 are

$$\frac{1}{D}q^{2\beta} - \frac{3}{2I^2}q^2 + \frac{I_1}{D^{\frac{5}{2}}}\frac{3}{2}u_2 q^{5\alpha-1} + \frac{I_3 V}{D^{\frac{3}{2}}}q^{3\alpha-1} - \frac{I_4 V\kappa^2}{D^{\frac{1}{2}}}q^{\alpha-1} + ... = 0 \tag{19}$$

At $\kappa \to 0$, the dominant balance for small q (big |s-s'|) gives

$$3\alpha - 1 = 2, \ D = \left(\frac{2}{3}I^2 VI_3\right)^{\frac{2}{3}} \tag{20}$$

or $\alpha=1$. In the absence of screening, a full extension of the polyion segment is predicted. A moderately screened polyion can be viewed as a self-avoiding chain over distances considerably exceeding the screening length $1/\kappa$ while its segments behave as rigid rods over the range short in comparison with $1/\kappa$. In poor solvents, the three-particle term of equation 2 represents an important short-ranged repulsive contribution. In the absence of the screening, analogous considerations as applied above reveal a fully stretched long-distance configuration, $\alpha \sim 1$. For smaller separations $|s-s'|$, a rather different scaling may be observed. Here, the dominant balance has to be applied to the large q regime. Depending on the sign and the magnitude of the short-ranged coupling u_2, this may yield the scaling exponent $\alpha=1/2$ or $\alpha=3/5$. A globally extended segment can comprise smaller blobs $(2,11,17)$ with internal scaling of a gaussian or a self avoiding coil. In poor solvents, characterized by negative values of the parameter u_2, attractive forces between the units would lead to the formation of collapsed blobs with $\alpha=1/3$. A detailed analysis along these lines will be reported (35).

Numerical Results

The system of L/2 nonlinear equations 12 for coefficients g(q) is solved by direct iteration. The procedure is initiated by assuming an ideal Gaussian behavior of the ring and using the corresponding mean square distances $<|\mathbf{r}(s)-\mathbf{r}(s')|^2>$ in calculation of corrected g(q). From those, improved distances $<|\mathbf{r}(s)-\mathbf{r}(s')|^2>$ are determined and the procedure is repeated until a self consistent solution of a sufficient accuracy is obtained. The use of mixing parameter of rather low value between 10^{-3} and 10^{-1} is needed to ascertain the convergence of the iteration. The number of iterations needed depends on the length of the chain L, the charge q and the screening parameter κ. About 10^2-10^3 cycles were usually sufficient to obtain the accuracy better than 0.01% in calculated mean square distances $<|\mathbf{r}(s)-\mathbf{r}(s')|^2>$. In the following paragraph, we present numerical results for polyion expansion obtained from equations 12-14 at various conditions. The model parameters in these calculations correspond to aqueous solutions of vinylic polyions of monomer length l=2.52 A and the degree of ionization determined by Manning's limiting charge density of the polyion $(40\text{-}42)$. According to the concept of ion condensation $(40\text{-}42)$, seen to conform with the results of computer simulation (43), these results should be approximately valid for all ionizations above the critical value. The monomer units are treated as hard spheres of diameter l. In Table I, the mean square distances $<|\mathbf{r}(s)-\mathbf{r}(s')|^2>$ and the scaling exponents $2\alpha= d\ln<|\mathbf{r}(s)-\mathbf{r}(s')|^2>/d\ln$ L for various degrees of polymerization L and the ionic strengths are collected. At very strong screening, numerical data approach the results for self avoiding cyclic polymers $(32,33)$. As pointed out by des Cloizeaux $(32\text{-}34)$, the self-avoiding rings under present approximations are characterized by the Reiss exponent $2\alpha=4/3$ and this is confirmed by the present results. The nonscreened polyions, on the other hand, behave as fully extended rings. In a few extreme cases, α slightly exceeds unity reflecting not only the full expansion but also the extension of individual links due to the increase of the electrostatic repulsion with growing L. The problem is not seen in studies of intramolecular exponent measured at constant L. Figure 1 illustrates the dependence of the renormalized intramolecular scaling exponent $2\alpha' = \{d\ln<|r(s)-r(s')|^2>/d\ln|s-s'|\}/\{(\pi|s-s'|/L)\cot g(\pi|s-s'|/L)\}$ on the relative distance $|s-s'|$. These results are corrected in order to eliminate the effect of the mean curvature present in finite-sized polymer rings.

Table I Scaling exponents $2\alpha = d\ln<|r(s)-r(s')|^2>/d\ln L$, renormalized exponents $2\alpha'(L/2)$, and mean square distances $\Delta R^2 = <|r(s)-r(s')|^2>$ in units 1 for various degrees of polymerization L and ionic strengths I [mol dm^{-3}]

I	0.001	0.01	0.1	1.0
N=100	2.08	1.94	1.69	1.46
	1.84	1.79	1.65	1.42
	27.6	25.8	21.4	16.3
200	2.04	1.87	1.64	1.43
	1.90	1.83	1.64	1.40
	56.4	49.8	38.0	26.9
500	1.96	1.77	1.56	1.40
	1.93	1.83	1.59	1.37
	141	115.2	79.1	51.4
2000	1.88	1.67	1.47	1.38
	1.91	1.73	1.48	1.32
	531	384	226	134.7

Weakly screened polyion is globally extended over large distances, the scaling exponent decaying with the increase in the ionic strength of the solution. Decreasing distance $|s-s'|$ and concomitant weakening of the screening of Coulombic forces lead to a slow increase in α' until the curves pass the maxima located at relatively small values $|s-s'|<<L$. A rapid drop in α' seen at even smaller $|s-s'|$ reveals the existence of electrostatic blobs predicted in earlier analysis. A full discussion of these issues can be found in a related report (35).

The methods of the present work are not restricted to the particular polymer Hamiltonian and can be used in cases with quite arbitrary distribution of interacting groups along the chain. This invites studies of complex macromolecules and phenomena like helix-coil transition, folding or collapse of the polymer in solution. For illustration, we present the results of an analogous calculation for a cyclic polymer with excluded volume interactions and attractive forces that impose the formation of a helix-like configuration. Apart from the steric term of equation 1, the Hamiltonian includes the attractive contribution of the form $u_a(s,s')=(1/3) <|r(s)-r(s')|^2>\delta_{|s-s'|,6}$. Figure 2

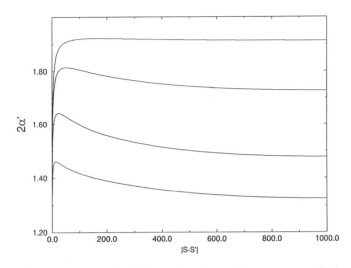

Figure 1 The renormalized intramolecular scaling exponent $2\alpha'$ as a function of the separation $|s\text{-}s'|$ at L=2000 and at ionic strengths (top to bottom): $I=10^{-3}$, 10^{-2}, 10^{-1} or 1.0 mol dm^{-3}.

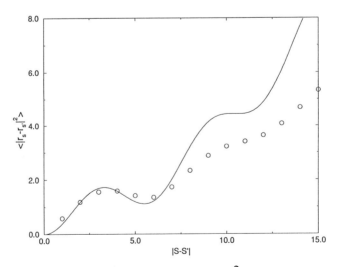

Figure 2 The mean square distance $<|\mathbf{r}(s)\text{-}\mathbf{r}(s')|^2>$ in a flexible polymer with helix-imposing intramolecular forces at L=192 (O) compared to a rigid helical configuration (——).

compares preliminary results for a flexible chain with above short-ranged interactions to those predicted by assuming a rigid helical configuration. Other examples and a generalization of present perturbative theory to nonisotropic situations such as the macromolecule in external field or under tension (*44*) will be described at a later date.

Literature Cited

1. Kuhn, W.; Kunzle, O.; Katchalsky, A. *Helv. Chim. Acta* **1948**, *31*, 1994-2037.
2. DeGennes, P. G.; Pincus, P.; Velasco, R. M.; Brochard, F. *J. Phys.* **1976**, *18*, 1461-1473.
3. Skolnick, J.; Fixman, M. *Macromolecules* **1977**, *10*, 944-948.
4. Bailey, M. *Macromolecules* **1977**, *10*, 725-730.
5. Odijk, T. *J. Polym. Sci., Polym. Phys. Ed.* **1977**, *15*, 477-483.
6. Odijk, T.; Houwart, A. C. *J. Polym. Sci., Polym. Phys. Ed.* **1978**, *16*, 627-639.
7. Fixman, M.; Skolnick, J. *Macromolecules* **1978**, *11*, 863-867.
8. Woodward, C. E.; Jonsson, B. *Chem. Phys.* **1991**, *155*, 207-219.
9. Granfeldt, M.; Jonsson, B.; Woodward, C. E. *J. Phys. Chem.* **1992**, *96*, 10080-10086.
10. Fixman, M. *J. Chem. Phys.* **1982**, *76*, 6346-6353.
11. Khoklov, A. R.; Khachaturian, K. A. *Polymer* **1982**, *23*, 1742-1750.
12. Kholodenko, A. L.; Freed, K. F. *J. Chem. Phys.* **1983**, *78*, 7412-7428.
13. Bawendi, M. L.; Freed, K. F. *J. Chem. Phys.* **1986**, *48*, 449-464.
14. Muthukumar, M. *J. Chem. Phys.* **1987**, *86*, 7230-7235.
15. Qian, C.; Kholodenko, A. L. *J. Chem. Phys.* **1988**, *89*, 2301-2311.
16. Qian, C.; Kholodenko, A. L. *J. Chem. Phys.* **1988**, *89*, 5273-5279.
17. Higgs, P. E.; Raphael, E. *J. Phys. I* **1991**, *1*,1-7.
18. Podgornik, R. *J. Phys. Chem.* **1991**, *95*, 5249-5255.
19. Podgornik, R. *J. Phys. Chem.* **1992**, *96*, 884-896.
20. Victor, J. M. *J. Chem. Phys.***1991**, *95*, 600-605.
21. Bratko, D.; Dawson, K. A. *Polym. Prepr.* **1993**, *34*, 936-937.
22. Brender, C.; Lax, M. *J. Chem. Phys.* **1977**, *67*, 1785-1787.
23. Brender, C.; Lax, M. *J. Chem. Phys.* **1981**, *74*, 2659-2660.
24. Brender, C. *J. Chem. Phys.* **1990**, *92*, 4468-4472.
25. Carnie, S. L.; Christos, G. A.; Creamer, T. P. *J. Chem. Phys.* **1988**, *89*, 6484-6496.
26. Christos, G. A.; Carnie, S. L. *J. Chem. Phys.* **1989**, *91*, 439-453.
27. Christos, G. A.; Carnie, S. L. *J. Chem. Phys.* **1990**, *92*, 7661-7677.
28. Reed, C. E.; Reed, W. F. *J. Chem. Phys.* **1991**, *94*, 8479-8486.
29. Reed, C. E.; Reed, W. F. *J. Chem. Phys.* **1992**, *96*, 1609-1620.
30. Valleau, J. P. *Chem. Phys.* **1989**, *129*, 163-175.
31. Doi, M.; Edwards, S. F. *The Theory of Polymer Dynamics,* Oxford Science Publishers, New York, NY, 1992.
32. des Cloizeaux, J. *J. Phys. Soc. Japan* **1969**, *26*, 42-45.

33. des Cloizeaux, J. *J. Phys* **1970**, *31*, 715-736.

34. des Cloizeaux, J.; Jannink, G. *Polymers in Solution. Their Modelling and Structure,* Clarendon Press, Oxford, 1990.

35. Bratko, D.; Dawson, K. A., submitted.

36. Barrat, J. L.; Boyer, D., submitted.

37. Edwards, S. F.; Singh, P. *J. Chem. Soc., Faraday II* **1979**, *75*, 1001-1029.

38. Arfken, G. *Mathematical Methods for Physicists*, Academic Press, San Diego, CA, 1985.

39. Bender, C. M.; Orszag, S. A. *Advanced Mathematical Methods for Scientists and Engineers,* McGraw-Hill, New York, NY, 1978.

40. Manning, G. S. *J. Chem. Phys.* **1969**, *51*, 924-933.

41. Manning, G. S. *Q. Rev. Biophys.* **1978**, *11*, 179-246.

42. Manning, G. S. *Acc. Chem. Res.* **1979**, *12*, 443-449.

43. Bratko, D.; Vlachy, V. *Chem. Phys. Lett.* **1985**, *115*, 294-298.

44. Dawson, K. A.; Bratko, D. In *Condensed Matter Theories,* Vol. 8, Blum, L.; Malik, F. B., Eds.; Plenum, New York, NY, 1993.

RECEIVED August 6, 1993

Chapter 4

Branched Polyelectrolytes

M. Daoud

Laboratoire Léon Brillouin, Centre d'Etude de Saclay, 91191 Gif-sur-Yvette, Cédex France

We discuss the conformation of randomly branched polyelectrolytes. Because these are synthesized in the neutral state, and then charged electrically, the distribution of molecular weights is the same as for the neutral polymers, and is assumed to be given by percolation. We review the various regimes that might be found when concentration is changed, in the absence of salt. We find the possible existence of two dilute regimes. In the first one electrostatic interactions are present, and the polymers are swollen. In the second one, the Debye-Hückel length is smaller than the radius of the largest macromolecules, and screening is present, leading to a neutral behavior for large distances. In the various regimes, we consider the structure of the polymers and the effects of polydispersity.

Randomly branched polymers have been studied recently both theoretically *(1-5)* and experimentally *(6-11)*. Polydispersity is very important *(12)*, and effective exponents *(12-14)* may be observed in the averaged quantities that are measured experimentally. The latter exponents are different from those of a single macromolecule, and are also related to the distribution of masses. All these studies were performed on electrically neutral systems. Polyelectrolytes were somewhat discarded, even though they constitute an important class of materials. Some experimental results *(15-17)* have recently appeared, but very few theoretical work was made *(18-19)*. This is especially true for the finite branched polymers constituting the sol part. It is however possible to consider these polymers within a simplified Flory approximation. The reason for this is that they are usually synthesized in the neutral state and then charged electrically. Because of this procedure, the distribution of masses should be very similar to the one that was observed in the neutral case. Therefore, the polydispersity effects that were observed in the latter case should also be present in this case. Moreover, because of the presence of the counterions, we know that screening effects are present in these electrically charged systems. In order to characterize them, one usually introduces a

Debye-Hückel length λ_D. The latter depends only on the density of free charges in the solution. As we will see, except for very dilute solutions, this screening length becomes smaller than the radius of the largest polymers, and therefore screening effects for the charges are present even in the dilute regime, that is below the overlap concentration C^*. This leads us to introduce a second concentration, C_S, smaller than

0097–6156/94/0548–0045$06.00/0

C*, and corresponding to the appearance of these screening effects. We are mainly interested in this effect. We stress that the following is a first approach to this problem. Many approximations are made in order to focus on the possibility of having two dilute regimes instead of only one usually. As far as we can see, the main approximations concern the value of the exponents, which are assumed here to have the Flory values. We also assume that the counterions are free. This is obviously a rough approximation, as the attractive potential by larger and larger polymers will bring the counterions back to the polymer. Thus we assume that the various macromolecule have a constant fraction f of their monomers that is charged. Finally, for small f, we also neglect the possibility that uncharged parts may phase separate because the solvent is not good for them.

In the following, we will first recall briefly the properties of the distribution of molecular weights. We will assume that the percolation distribution is valid, and that there is a power law decay for increasing masses. We will then turn to the swelling properties of a single branched polyelectrolyte and to the effects of the polydispersity in the case of very dilute solutions. Section 4 deals with the eventual existence of screening of the electrostatic interactions above a concentration C_S. Section 5 is devoted to the concentration effects. We will introduce the screening and overlap concentrations C_S and C*, and discuss the conformation of the polymers in the various regimes that will be defined. All the following will be in the Flory approximation. Such approximation usually gives values for the exponents that are very close to the best known ones calculated by computer simulations or renormalization group, but one has to be very cautious about it because we do not really control the validity of this approximation. Nevertheless, what is important in our opinion is the eventual existence of a dilute regime where power law dependences with concentration may be observed for many quantities. As we will see, the existence of such regime does not depend crucially on the exact values of the exponents.

The distribution of masses.

We consider multifunctional units that react with each other to form randomly branched polymers. We assume that in the reaction bath there is no solvent, and that the monomers are not ionized. The ionization process is assumed to take place after the polymers are formed, and the synthesis has been quenched. Because we assume that we are dealing with systems at equilibrium, the case when solvent and some ionization of the monomers are present is in principle equivalent to a solution that has been synthesized without any solvent, and then add the solvent and /or adjust the pH of the solution. There results a semi-dilute solution, to be studied below. Thus we consider the synthesis only in the neutral case. The latter was studied carefully these last few years, and it seems that percolation provides a good description for the sol-gel transition (20-21) for a wide variety of cross-linking procedures (22-24). In what follows, we will consider only this case, and assume that it describes the formation of branched polymers. The main result is that there is a very wide distribution of masses, characterized by the probability $P(N,\varepsilon)$ of finding a macromolecule made of N segments at a distance $\varepsilon = p - p_c$ from the gelation threshold. Here p is the extent of reaction for instance. This function decreases (25) as a power law, with a cut-off at large masses that also depends on ε.

$$P(N,\varepsilon) \sim N^{-\tau} p(N/N_z) \tag{1}$$

where τ is an exponent that is related to the fractal dimension of the polymers in the reaction bath. The second moment of the distribution is the weight average molecular weight, and also diverges at the gelation threshold. It was shown that it may be related to the largest mass N_Z

$$N_W \sim N_Z^{(2D-3)/D} \approx N_Z^{4/5} \tag{2}$$

where D is the fractal dimension of every macromolecule in the reaction bath, and 3 is the dimension of space. The exponent in the right hand side is given within the Flory approximation for D. The radius R_Z of the largest cluster may be related to its mass via the fractal dimension

$$N_Z \sim R_Z^{D} \tag{3}$$

We assume that equation 3 is not valid only for the largest macromolecule, but relates the radius to the mass whatever the polymer in the distribution.

Finally, the exponent τ of the distribution $P(N,\varepsilon)$ was also related to D. It was found

$$\tau = 1 + 3/D \tag{4}$$

The latter equation was interpreted *(26)* as a generalized C* situation. It implies that if one defines classes of polymers with given mass, each of these classes is at C*, that is there is a filling of space by the macromolecules of similar sizes. Because they are fractal however, there is space left, which is filled by smaller polymers, with the prescription that every class is space filling. When all classes have been considered, no space is left. Let us note that as the gelation threshold is approached, the characteristic masses diverge in different ways, so that polydispersity also diverges. This has to be opposed to other models where there is only one characteristic diverging mass *(27)*. In what follows, we will assume that this distribution of masses is given and fixed. The various manipulations that are subsequently done on the polymers, such as dilution or adding salt, etc..., are made with a fixed distribution, and only the conformation of the polymers changes, but their proportions remains constant.

Very dilute solutions. Neutral blobs.

Once the synthesis is made, we add solvent in excess, so that the resulting solution is dilute. We assume that during this dilution the charges appear on the macromolecules, or are created, by changing the pH of the solution for instance, so that the result is a dilute solution of polyelectrolytes. We also assume that a constant fraction f of the monomers of all polymers is charged. It is clear that because both the swelling and the electrostatic screening effects are closely related to the density of charges, the effect of the latter is important. We consider first the case of a single polyelectrolytes, within a Flory approximation. Following Isaacson and Lubensky *(28)*, The free energy may be written as

$$F = \frac{R^2}{N^{1/2} a^2} + f^2 N^2 \left\{ \frac{\lambda_B}{R} \right\}^{d-2} \tag{5}$$

where the first term is an elastic contribution, the second one the electrostatic one, λ_B is the Bjerrum length *(35)*, and is assumed to be of the same order than the step length a, d is the dimension of space, and we assumed the Zimm-Stockmayer *(29)* result for an ideal branched polymer:

$$R_O \sim N^{1/4} a \qquad (6)$$

Minimizing equation 5 with respect to R gives

$$R \sim N^{5/2d} [\frac{af}{\lambda_B}]^{2/d} \lambda_B \qquad (7)$$

and a fractal dimension D_O

$$N \sim R^{D_O} \qquad (8a)$$

with

$$D_O = 2d/5 \qquad (8b)$$

In our case, we get $D_O = 6/5$. Note that one recovers the ideal Zimm Stockmayer value for $d = 10$, so that large deviations are expected for the three dimensional case we are considering. Note also that in order to get equation 7, we neglected the excluded volume interactions in the free energy, equation 5. As a result, there is a condition that we have to set on the resulting radius, namely that it has to be larger than that of a neutral branched polymer with equivalent mass. From previous studies, we know the latter is

$$R_n (N) \sim N^{D_a} a \qquad (9a)$$

with

$$D_a \sim 2(d+2)/5 \qquad (9b)$$

This leads us to compare these estimates. For $d=3$, we find that the electrostatic effects are important if the polymer is sufficiently large or charged. Equation 8 is valid only if $N > N^*$ for given f (or $f>f^*$ for given N) such that

$$N^* \sim f^{-2} \frac{a}{\lambda_B} \qquad (10a)$$

This in turn allows us to introduce local blobs, with neutral behavior. These are made of N^* units and have a radius R^* related to N^* by equation 9:

$$R^* \sim N^{*1/2} a \qquad (10b)$$

For larger distances, the behavior of the polymer is governed by equation 8 if one takes the blob as a statistical unit. Thus for $d=3$, we have:

$$R \sim [\frac{N}{N^*}]^{5/6} R^* \qquad (10)$$

Using equations 9 and 10, one recovers equation 8. We finally observe that in the above analysis, we assumed that the polymer is in a good solvent. This might be questioned because the solvent may be a poor solvent for the uncharged parts of the polymer. In the following, we will not consider this local structure, and rather focus on the mass dependence of the radius.

The latter result may be observed only if one has a single molecular weight. This implies a fractionation of the distribution. If the whole distribution is kept, averages have to be made. These lead to effective fractal dimension as we will see now. The average radius that is measured by light or neutron scattering is

$$R_z^2 = \frac{1}{N_w} \int dN \ N^2 \ P(N,\varepsilon) \ R^2(N) \tag{11}$$

Using equations (1),(4), (7) and (8), and eliminating ε with equation(3), we get

$$R_z^2 \sim N_w^{25/12} \ [\frac{af}{\lambda_B}]^{4/3} \ \lambda_B^2 \tag{12}$$

Another measurable quantity, namely the intrinsic viscosity may also be measured. For one mass, this would be

$$[\eta(N)] \sim N^{-1+3/D_0} \approx N^{3/2}[af]^2 \ \lambda_B \tag{13}$$

where we used equation 8. The measured quantity however is an average over the whole distribution. It was shown *(30)* that this is

$$<[\eta(N)]> = \int^{N_z} dN \ N \ P(N) \ [\eta(N)] \tag{14}$$

Performing the integration and eliminating N_z with equation 2, we get

$$<[\eta(N)]> \sim N_w^{1-\tau+3/D_0} \approx N_w^{13/8} \ [af]^2 \ \lambda_B \tag{15}$$

One might also calculate the overlap concentration C^* where the various macromolecules start overlapping each other. However, an important concentration effect appears even below C^*. This is related to the screening of the electrostatic interactions.

Screening of the electrostatic interactions.

The previous results are valid for very dilute solutions. However, even in this regime, when the monomer concentration C is increased, some screening of the electrostatic interactions appears, because of the presence of the free counterions in the solution. These effects are not important as long as the concentration of the solution is small. However, because of the polydispersity and the compactness of the polymers, they become important sooner for the branched polymers than for the linear ones. Whereas they become important in the semi-dilute regime for linear polyelectrolyte, we will see that they are already even in a dilute range for the branched case. Screening effects are usually described by a Debye-Hückel screening length λ_D, related to the presence of free counterions in the solution. This length depends on concentration. As mentioned in the introduction, we assume that all counterions are free. Therefore their concentration is fC, where C is the monomer concentration. The Debye length is thus

$$\lambda_D \sim (fC\lambda_B)^{-1/2} \tag{16}$$

and is smaller than the radius of the polymers as long as the concentration is low enough. There is a special concentration C_S where the screening length becomes of the order of the radius of the largest polymers, with size N_Z. Using equations 7 and 16, we get

$$C_S a^3 \sim N_W^{-25/12} f^{-7/3} [\frac{\lambda_B}{a}]^{-5/3} \tag{17}$$

Therefore, for concentrations smaller than C_S, screening effects may be considered as perturbations to the laws derived in the previous section . For concentrations larger than C_S however, these effects are no longer corrections, but become predominant. For any concentration C larger than C_S, the largest clusters are screened, and one has to partition the distribution of masses into two parts, and to introduce a cross-over mass g_S between two behaviors. The larger masses are screened whereas the smaller ones are still interacting with long range electrostatic interactions. Generalizing equation 17, we get the cross-over mass

$$g_S \sim (C a^3)^{-3/5} f^{-7/5} [\frac{a}{\lambda_B}] \tag{18}$$

Note that this is the largest unscreened mass, and that , for the lower mass part of the distribution, the corresponding weight average molecular weight g_{SW} is $g_{SW} \sim g_S^{4/5}$. Using equation 18 together with 12 gives the size λ_D of the screening length, equation 16, as expected. Thus above C_S, the conformation of the polymers with mass smaller than g_S is the same as discussed above. The structure of the larger polymers is obtained by introducing blobs made of g_S monomers, with size λ_D. Using such blobs as statistical units, one may neglect the electrostatic interactions between blobs as a first approximation. Therefore, the conformation of the largest masses in the distribution is the same as in the neutral case. This was studied previously and, for dilute solution, one gets a fractal dimension D_a, which, in a Flory approximation, is given by equation 9b. Therefore the radius of a large polymer is, for d=3:

$$R(N) \sim \{ \frac{N}{g_S} \}^{1/2} \lambda_D \tag{19a}$$

Using equations 8, 16 and 18, we get, in the three dimensional case

$$R(N) \sim N^{1/2} (C a^3)^{-1/5} f^{1/5} a \qquad (N \gg g_S) \tag{19b}$$

As in the neutral case, the observed (average) radius is

$$R_Z^2 = \frac{1}{N_W} \int_{g_S}^{N_Z} N^2 R(N)^2 P(N) \, dN \tag{20}$$

Neglecting the lower limit of the integral for $C \gg C_S$, and using equation 2 to eliminate N_z in favor of the measurable N_w, we get

$$R_z \sim N_w^{5/8} (C a^3)^{-1/5} f^{1/5} a \qquad (21)$$

Note that in both equations 19b and 21, the Bjerrum length has disappeared, so that there is no longer any explicit reference to the electrostatic interactions. It is interesting to consider the reduced viscosity $\eta_r = (\eta - \eta_s)/C \eta_s$ in this concentration regime. Here η and η_s are respectively the viscosities of the solution and the solvent. The contribution of each mass is

$$\eta_r(N) \sim R(N)^3/N$$

$$\sim N^{1/2} [Ca^3]^{-3/5} f^{3/5} a^3 \qquad (22)$$

Therefore the average relative viscosity is

$$<\eta_r> \sim \int^{N_z} dN \ N \ P(N) \ \eta_r(N) \qquad (23)$$

leading to

$$<\eta_r> \sim N_w^{3/8} [Ca^3]^{-3/5} f^{3/5} a^3 \qquad (24)$$

Thus we find that the relative viscosity of a dilute solution of branched polyelectrolytes decreases as concentration increases. Such result is in agreement with the Fuoss law, valid for linear polyelectrolytes. The latter was recently discussed by Witten and Pincus *(31)* who found a very different reason for this behavior in the case of linear charged chains.

We conclude this section by noting that the existence of this unusual dilute regime, $C_S \ll C \ll C^*$, that has concentration dependences that are not perturbations is related to the branched nature of the polymers. At the overlap concentration C^*, the largest polymers are therefore screened and behave as neutral branched macromolecules. The latter concentration corresponds to space filling by the polymers. This may be written in the following form:

$$\{C^* a^3\}^{-1} \sim \int^{N_z} R(N)^3 \ P(N,\varepsilon) \ dN \qquad (25)$$

In principle, this should be split into two parts, corresponding to the unscreened and screened portions of the distribution of masses. In the following, we will neglect the contribution of the smallest polymers, and consider only the largest ones. Using equations 1, 2, and 21, we get

$$C^* a^3 \sim N_w^{-15/16} f^{-3/2} \qquad (26)$$

Note that although this concentration is small, it is not as small as for linear polyelectrolytes because of these screening effects.

For higher concentrations, $C \gg C^*$, the polymers start interpenetrating each other. This is studied in next section.

The semi-dilute regime.

The overlap concentration C^* is the highest concentration where the polymers do not overlap and where one may assume that they behave independently of each other. We turn now to more concentrated solutions, well above C^*, where scaling may be applied. In this concentration range, two lengths are present, namely the Debye length λ_D, equation 16, and a screening length ξ corresponding to the interpenetration of the large polymers by the smaller ones. Before we calculate this length, we describe briefly the conformation of the polymers. This is identical to what was found for neutral branched macromolecules *(32)*. One has to split again the distribution of masses into two parts. The smaller part, for masses less than g, corresponds to swollen state, equation 21. These polymers behave as in a dilute solution. The polymers with size larger than g are penetrated by the smaller polymers and are screened. Their overall behavior is therefore the same as in the reaction bath, that is they have the same fractal dimension as in a melt. We know that the latter is the fractal dimension D of percolation. We define the blob ξ as the size of the cross-over mass g. As for neutral polymers, it is a local property that has to be independent of the size N_z of the largest polymers, and should depend on concentration only. One may estimate it with a scaling argument. The characteristic distance has the scaled form

$$L \sim N_W^{5/8} \, [Ca^3]^{-1/5} \, f^{1/5} \, a \, g(\, C/C^*) \tag{27}$$

Assuming that f(x) behaves as a power law for x>>1, and using equation 29 and the condition that ξ is independent of N_W, we get

$$\xi \sim [Ca^3]^{-13/15} \, f^{-4/5} \, a \tag{28}$$

Note that $\xi \gg \lambda_D$, and that both lengths become of the same order for large concentrations. This may also be seen by comparing the number of monomers g_s in a Debye length and g in the screening length. The latter may be obtained by scaling arguments, with the limit that $g \approx N_z$ for $C \approx C^*$. One finds

$$g \sim [Ca^3]^{-3/4} \tag{29a}$$

Note that this is the largest mass in the distribution of small clusters, and that it corresponds to a weight average mass g_W

$$g_W \sim [Ca^3]^{-16/15} \tag{29b}$$

Finally, the average radius of gyration in the semi-dilute regime may also be obtained by scaling arguments similar to those we just used for the screening length ξ. The constraint now is that the fractal dimension of each macromolecule is the one of percolation Therefore, the average radius has a mass dependence that is proportional to $N_W^{1/2}$. We find

$$R_Z \sim N_W^{1/2} [Ca^3]^{-1/3} a \tag{30}$$

Note that this corresponds to an overall behavior for each of the polymers   •

$$R(N) \sim N^{2/5} [Ca^3]^{-1/3} a \tag{31}$$

The concentration dependence both in equations 29 and 30 is indicative of a local behavior that is different from the overall behavior. Indeed, as discussed above, one may introduce blobs made of g monomers, and size ξ such that for instance

$$R(N) \sim \{N/g\}^{2/5} \xi \tag{32}$$

Using equation 29a and generalizing equation 19b to the blob, one recovers equation 31. A similar treatment is also valid for the average radius. In both cases, this amounts to say that the behavior of large polymers is the same as that of the percolation clusters, if one takes the blob as a statistical unit. These laws may be checked by light or neutron scattering experiments. As for linear polyelectrolytes however, these are not straightforward to interpret. The reason for this is that the scattered intensity at very low scattering vectors is related to the osmotic compressibility. The latter is itself related to the fact that the osmotic pressure does not depend on the polymer-polymer contacts, but rather to the concentration in counterions. Therefore, we expect, as in the linear case, that the intensity is very low for q going to zero, and exhibits a maximum.

The concentrated regime.

As monomer concentration is still increased, a last cross-over occurs, to a "dense" regime where the electrostatic interactions are screened at all distances, and the behavior of the polymers is the same as that of neutral ones. In the semi-dilute regime, we had three different distance scales. For small distances, because electrostatic interaction is not strong enough, there is a neutral a neutral blob with size R^*. For larger distances the electrostatic interactions come into play, for distances smaller than the Debye screening length λ_D. Finally, for distances larger than λ_D, the long range forces are screened, and the behavior of neutral branched polymers is recovered, although the size of the concentration blobs ξ is different from what was obtained in the neutral case. All these distances depend on monomer concentration. The cross-over from the semi-dilute regime to the concentrated one corresponds to the special concentration C^{**} when the sizes of neutral blob R^* and the screening length λ_D become equal. Using equations 10b and 16, we get:

$$C^{**} a^3 \sim f \tag{33}$$

For concentrations larger than C^{**}, or for a given value of C, for small values of the fraction f of charged monomers, the electrostatic interactions are no longer playing any role in the problem. Thus we fully recover the neutral case. This implies that the correlation length ξ, equation 28, corresponding to the concentration blob should no longer depend on the f or λ_B. This is obtained by using the scaling equation for ξ: In order to describe the cross-over from semi-dilute to dense regimes, we assume that

$$\xi \sim [Ca^3]^{-13/15} f^{-4/5} a\ g\ (C/C**) \tag{34}$$

Assuming that g(x) behaves as a power law for large x, in the dense region and insisting that ξ should be independent on f, we get

$$\xi \sim [Ca^3]^{-5/3} a \quad (C >> C**) \tag{35}$$

Equation 35 is identical to what was obtained in the semi-dilute regime in the case of neutral polymers. Note also that it may be obtained directly from equations 9 and 10: we saw in the first section that small polymers have neutral behavior. The overlap concentration C* between dilute and semi-dilute regimes is evaluated directly in this case using equations 9 and 25. As for neutral polymers, we find

$$C* \sim N_z^{-3/10} \sim N_w^{-3/8} \tag{36}$$

This corresponds to the cross-over line between regions DN and SDN in figure 1. Finally, note that we gave the analysis with neutral blobs only when screening was not present, region I in figure 1. It is clear that such analysis may be extended to the case when screening is present. The cross-over from the regime where screened electrostatic interactions are present, region II in figure 1, to the neutral one is then obtained by comparing the screening length with the size of the neutral blob. As this is precisely what we did when we considered the cross-over from the semi-dilute to the dense regimes, such comparison leads us to the C** line that we considered above.The resulting diagram shown in figure 1, summarizes the various regimes that we considered.

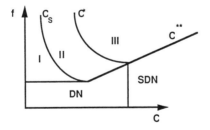

Figure 1. Crossovers between regions with increasing concentration.

Conclusion

We considered the conformation of randomly branched polyelectrolytes within a Flory approximation. It is well-known that the validity of such approach has to be checked. It is important to note at this level that de Gennes *(33)* very recently made a conjecture contradicting the Isaacson-Lubensky results, equation 8. His argument is that in the ideal case, the radius of the polymer is as if we had a linear chain made of $N^{1/2}$ monomers. This would imply that the conformation would already be stretched in the excluded volume case, because the radius is already proportional to $N^{1/2}$. Therefore, a variation faster than this is not to be expected, and one gets for the branched polyelectrolyte the same behavior instead of equation 8. This of course would change completely our results, and very few would survive such a criticism. The idea that another concentration is present would be among these. Indeed, because the Debye length λ_D varies as $[Cf\lambda_B]^{-1/2}$, there is a concentration C_s' such that the radius of the largest polymers is of the order of λ_D. We would then find

$$[C_s'f\lambda_B] \sim N_z^{-1} \sim N_W^{-5/4} \tag{37}$$

Because the difference in the radii of the polymers between the regimes above and below C_s' is merely in constants however, it is not clear to us at this point what the differences are between these concentration ranges as long as the conformation of the macromolecules is concerned. It would certainly be interesting to check this point, for instance with computer simulations, and to look experimentally at the effect of the density of charges on the structure of branched polymers *(34 , 35)*.
If we accept the validity of the Flory approximation, equation 8, however, charge effects would be important, and the conformation of the polymers would much more expanded below C_s than above. The screening effects would be present in the dilute regime, because $C_s \sim N_W^{-25/12}$ is smaller than C^*, and lead to important changes between these two concentration regimes. More precisely, we found that explicit concentration dependences would be present for all measurable quantities in the dilute and screened regime II of figure 1. Among others, we found that the relative viscosity would decrease as concentration increases between C_S and C^*. An important result is that the dilute regimes should be easier to reach in the case of branched polymers whereas they were much too dilute for linear chains *(37)*. We found $C^*a^3 \sim N_z^{-3/4} \sim N_W^{-15/16}$, instead of $C^* a^3 \sim N_z^{-2}$ in the linear case. This would allow for instance a search for an eventual regime where ordering of the polymers would be present The latter question was not addressed here. Another interesting concept that we found is the eventual presence of neutral blobs when the fraction f of charged monomers is low. In this paper, we considered the simplest possibility for these blobs, namely that they behave as neutral branched polymers in a good solvent. This assumption has obviously to be studied more carefully by including the option that there might be locally a poor solvent *(37)*

References.

1. Stockmayer, W. H., *J. Chem.Phys.* **1943**, *11*, 45.
2. De Gennes, P. G., Scaling Concepts in Polymer Physics, Cornell University Press,Ithaca, **1979**.
3. Flory, P. J., Principles of Polymer Chemistry, Cornell University press, Ithaca **1953**.
4. Benedek, G. B., "*The Theory of the Sol-Gel Transition,*" lecture at E.T.H. Zurich **1980**.
5. Daoud, M., Martin, J.E., in *The fractal approach to Heterogeneous Chemistry* , D. Avnir Ed. , J. Wiley , **1990**.
6. Candau, S. J., Ankrim, M., Munch, J. P., Rempp, P., Hild, G., Osaka, R., in *Physical Optics of Dynamical Phenomena in Macromolecular Systems*, W. De Gruyter,Berlin, **1985**, 145.
7. Adam, M., Delsanti, M., Munch, J. P., Durand, D , *J. Physique* **1987**, *48* , 1809.

8. Allain, C., Salome, L., *Macromolecules* **1987**, *20* , 2957.
9. Burchard, W., *Adv. Pol. Sci.* **1983**, *48* , 1.
10. Schosseler, F., Leibler, L., *Macromolecules* **1985**, *18* , 398.
11. Patton, E. V., Wesson, J.A., Rubinstein, M., Wilson, J. C., Oppenheimer, L. E. *Macromolecules* **1989**, *22* , 1946.
12. Bouchaud, E., Delsanti, M., Adam, M., Daoud, M., Durand, D., *J. Physique Lett.* **1986**, *47* , 1273.
13. Martin, J. E., Ackerson, B. J., *Phys. Rev.* **1985**, *A31*, 1180.
14. Daoud, M., Family, F., Jannink, G., *J. de Phys. Lett.* **1984**, *45* , 119.
15. Moussaid, A., Munch, J.P., Schosseler, F., Candau, S.J., *J. Physique II* **1991**, *1*, 637.
16. Schosseler, F., Moussaid, A., Munch, J.P., Candau, S.J., *J. Physique II* **1991**, *1,* 1197.
17. Schosseler, F., Ilmain, F., Candau, S.J., *Macromolecules* **1991**, *24*, 225.
18. Borue, V., Erukhimovich, I. , *Macromolecules* **1988**, *21*, 3240.
19. Joanny, J.F., Leibler, L., *J. Physique* **1990**, *51*, 545.
20. De Gennes, P. G., *J. Physique Lett.* **1976**, *37*, 1.
21. Stauffer, D., *J. Chem. Soc. Faraday Trans. II* **1976**, *72*, 1354.
22. Leibler, L., Schosseler, F., in *Physics of Finely Divided Matter*, Springer Proc. Phys. 5,135, N. Boccara and M. Daoud eds., Springer Verlag,Berlin **1985**.
23. Adam, M., Delsanti, M., Durand, D., *Macromolecules* **1985**, *18*, 2285.
24. Lapp, A., Leibler, L., Schosseler, F., Strazielle, C., *Macromolecules* **1989**, *22*, 2871.
25. Stauffer, D., Introduction to percolation theory , Taylor and Francis **1985**.
26. Cates, M. E., *J. de Phys. Lett.* **1985**, *38* , 2957.
27. Meakin, P., in *The fractal approach to Heterogeneous Chemistry* , D. Avnir Ed. J. Wiley **1990**.
28. Isaacson, J., Lubensky, T. C., *J. Physique* **1981**, *42* , 175.
29. Zimm, B. H., Stockmayer, W. H., *J. Chem. Phys.* **1949**, *17* , 1301.
30. Sievers, D., *J. Physique* **1980**, *41*, L 535.
31. Witten, T.A., Pincus, F.Y., *Europhys. Lett* **1987**, *3*, 315
32. Daoud, M., Leibler, L., *Macromolecules* **1988**, *41*, 1497.
33. De Gennes, P.G., *Comptes Rendus Ac. Sci. (Paris) II* **1988**, *307*, 1497.
34. Odijk, T., *J. Pol. Sci. (Phys.)* **1977**, *15*, 477
35. Manning, G., *Quart. Rev. Biophys.* **1978**, *11*, 179.
36. De Gennes, P.G., Pincus, P., Velasco, R.M, Brochard, F., *J. de Physique* **1976**, *37*, 1461.
37. Khokhlov, A.R., *J. Phys.* **1980**, *A13*, 979.

RECEIVED September 7, 1993

Chapter 5

Molecular Dynamics Simulations of Charged Polymer Chains from Dilute to Semidilute Concentrations

Mark J. Stevens[1] and Kurt Kremer

Institut für Festkörperforschung, Forschungszentrum Jülich, Postfach 1913, D–5170 Jülich, Germany

We have performed molecular dynamics simulations on multichain systems of flexible charged polymers in which full Coulomb interactions of the monomers and counterions are treated explicitly. This model produces osmotic pressure and structure factor data that agrees excellently with experimental measurements. As expected we find the chain conformation changes from stretched to coiled as the density increases from dilute to semidilute values. However, the nature of the stretched conformations is different from theoretical predictions.

Polyelectrolytes remain one of the most mysterious states of condensed matter (*1–3*). This is in great contrast to the well developed theory of neutral polymer solutions (*4*). Experimentally bulk properties like the osmotic pressure (*2*) and the viscosity (*5*) are well known, but an understanding of the microscopic origin is lacking. This ignorance is especially critical since, for example, one of the prototypical polyelectrolytes is DNA which is of fundamental importance to biology. The lack of progress is in part due to unfortunate situation in which theory is best done in one regime (dilute) and experiment is best done in another regime (semidilute). We present results of simulations which bridge this gap and handle the inherent theoretical difficulties.

Polyelectrolytes pose difficulties not encountered in the theory of neutral polymers. The main difficulty is the long range nature of the Coulomb interaction which can only be handled theoretically by severe approximations. The usual approach is the Debye-Hückel approximation which treats the Coulomb pair interactions between monomers as strongly screened and thus short ranged. This approach should be valid at low concentrations. But, the experimentally relevant regime is at semidilute concentrations. Furthermore, the counterions enter only as a part of the screening, and any effect of the discrete nature of the counterions is neglected. Further theoretical complications arise because not only is a new length scale due to the Coulomb interaction introduced, but also other length scales occur due to the presence of extra species (counterions and any added salt). Attempts at scaling theories of polyelectrolytes are thus more speculative as one must choose the most relevant subset of the length scales.

The above difficulties suggest that simulations of polyelectrolytes are especially warranted. While several simulation studies have been done, they only considered a

---

[1]Current address: Corporate Research Science Laboratories, Exxon Research and Engineering Company, Annandale, NJ 08801

0097–6156/94/0548–0057$06.00/0

single chain (6). Yet, as noted above the experimentally relevant regime is at semidilute concentrations. Thus, it is essential to simulate a system of *several* chains with counterions explicitly treated. Here, we describe results of such molecular dynamics simulations on salt-free systems of polyelectrolytes that cover the concentration range from dilute to semidilute.

The rest of the paper is organized as follows. The simulation method and parameters are described. Then a brief discussion of relevant polyelectrolyte theory is given. Our simulation data is compared with experimental measurements of the osmotic pressure and the structure factor to verify accuracy. The chain structure is examined through a combination of the end-to-end distance, $\langle R \rangle$, and the radius of gyration, $\langle S^2 \rangle$. Finally, we discuss our calculated form factors.

Simulation Method

We use the freely-jointed bead-spring model of a polymer which has been very successful in simulations of neutral polymers (7,8). Each polymer chain consists of N_b monomers of mass m connected by a nonlinear bond potential. The bond potential is given by

$$U_{bond}(r) = -\tfrac{1}{2} k R_0^2 \ln(1 - r^2/R_0^2), \tag{1}$$

with $k = 7\epsilon/\sigma^2$ used for the spring constant, and with $R_0 = 2\sigma$ used for the maximum extent (8). Here, as throughout this paper Lennard-Jones units are used. Excluded volume between the monomers is included via a repulsive Lennard-Jones (RLJ) potential with the cutoff, r_c, at $2^{1/6}\sigma$:

$$U_{LJ}(r) = \begin{cases} 4\epsilon\left[\left(\dfrac{\sigma}{r}\right)^{12} - \left(\dfrac{\sigma}{r}\right)^6 - \left(\dfrac{\sigma}{r_c}\right)^{12} + \left(\dfrac{\sigma}{r_c}\right)^6\right]; & r \leq r_c \\ 0; & r > r_c \end{cases} \tag{2}$$

The counterions are given a repulsive core by using the same repulsive Lennard-Jones potential. Since simulations on neutral polymers give good results starting at 16 bead chains (7), the polyelectrolyte simulations were done with 16, 32 and 64 bead chains, and in some cases we extended the runs to $N_b = 128$ beads to test chain length dependence. The number of chains was either 8 or 16. The average bond length, $\langle b \rangle$, is 1.1σ.

Coulomb interactions pose special difficulties for simulations as well as theory (9). The long range interaction usually requires the Ewald summation method in order to include interactions of the periodic images. In our simulations the Coulomb interactions are evaluated by a spherical approximation to the Ewald sum given by Adams and Dubey (9). This approximation is better by an order of magnitude in the calculation of the energy than the minimum image evaluation which is the best used to date for polyelectrolytes (6). This approximation is good if the Coulomb interaction are not much stronger than the thermal interactions. The Coulomb strength is given by the Bjerrum length, $\lambda_B = e^2/\varepsilon k_B T$ which for this work is taken to be 1σ as many experiments and theoretical works have been done in this critical regime (1,3,10). For this value of λ_B the Coulomb pair interactions are generally less than $k_B T$.

The dynamics of the system is done at constant temperature, $T = 1.2\epsilon$, using the Langevin thermostat with damping constant $\Gamma = 1\tau^{-1}$, and timestep 0.015τ, where $\tau = \sigma(m/\epsilon)^{1/2}$ (7). Besides the necessary temperature control, the Langevin thermostat also stabilizes the trajectories making possible the long runs necessary. The length of the simulation is such that the chains move at least 10 times the contour length. For 64 bead chains this required about a million timesteps.

Theoretical Models

Polyelectrolytes are stretched in comparison to neutral polymers due to the Coulomb repulsion between charges on the polymer chain. Flory type arguments (*10–12*) give $R \sim N_b$ which corresponds to a rodlike conformation for the chain. Much theoretical work has been done calculating the effect of increasing the polymer concentration or adding salt on rigid rod state (*13–17*). These works describe the polymer as a wormlike chain with a persistence length, $L_p = L_e + L_o$, where L_o is the intrinsic persistence length of the uncharged polymer, and L_e is the electrostatic persistence length. The electrostatic contribution to the persistence length has been calculated to be asymptotically (*13,14*)

$$L_e = \lambda_{DH}^2 \lambda_B / 4 \langle b \rangle^2, \tag{3}$$

where $\lambda_{DH} = \kappa^{-1} = (4\pi \lambda_B \rho_m)^{-1/2}$ is the Debye length and ρ_m is the monomer (=counterion) density.

The polymer structure in the transition between dilute and semidilute concentrations is speculative (*1,3,10,18,19*). Odijk (*18*) and Hayter *et al.* (*1*) have proposed similar pictures of polyelectrolyte structure albeit for different physical reasons. The chain structure is that of an ideal chain of 'batonnets', where the 'batonnets' are straight segments of length, L_p. Using this picture Odijk has proposed a scaling theory and picture of polyelectrolyte conformation as a function of concentration (*18*). At dilute concentrations the chains are rodlike. With increasing concentration the chains remain rodlike until $L_e \approx L$ at monomer concentration $c_1^* = 1/16\pi \langle b \rangle^2 L$ ($\kappa L \gg 1$) where the chains begin to bend. A second threshold sets in when the strand-strand distance, $\xi > L_e$. This occurs at $c_2^* = 0.04/4\pi \lambda_B^3 = 0.003\sigma^{-3}$ which is independent of chain length. De Gennes *et al.* (*10*) have described an alternative picture in the semidilute regime. In their picture, the length of the straight segments is pinned at the average strand-strand distance.

Comparison of Simulation Results to Experiment

One quantity which is relatively easily measured and thus a good test of the simulations is the osmotic pressure, Π. The osmotic pressure has been measured on several systems and two scaling regimes are exhibited (*2*). At low densities $\Pi \sim \rho_m^{9/8}$. This dependence is predicted by Odijk's scaling theory (*18*). In the high concentration regime the scaling exponent changes to 9/4 which is the scaling for semidilute neutral polymers (*4*). These results suggest that the polyelectrolyte chains are stretched at dilute concentrations and are coiled at semidilute concentrations.

Our simulation results shown in Figure 1 give two scaling regimes. In the high density regime we find the same scaling exponent 9/4. For densities two orders of magnitude lower than the crossover density, $\rho_m^{(\Pi)} = 0.07 \pm 0.04$, our data is consistent with the 9/8 value. Our simulation extend the experimental results to lower densities where the noninteracting limit, $\Pi = k_B T \rho_m (1 + 1/N_b)$, is reached. At these lowest concentrations, the chain length dependence decreases.

One striking aspect of the osmotic pressure data is the chain length independence of the crossover density, $\rho_m^{(\Pi)}$. Since for $\rho_m > \rho_m^{(\Pi)}$ the Π-dependence is that of neutral polymers, the Coulomb interactions must be completely screened at $\rho_m^{(\Pi)}$. For complete screening, the Debye screening length should be less than all interparticle distances. The bond length is the shortest particle separation distance. We find $\lambda_{DH} = \langle b \rangle$ at $\rho_m = 1/(4\pi \lambda_B \langle b \rangle^2) = 0.066\sigma^{-3}$ for our set of parameters. This value agrees with our simulation results.

Experimentally, the polymer structure can be probed through measurements of the structure factor. The interpolymer structure factor has been measured and exhibits a peak whose position, q_m, varies with concentration, c, in a rather intriguing fashion

(20). At low concentrations, the peak position scales as $c^{-1/3}$ which corresponds to the variation of an average interpolymer separation. At high concentrations, the peak position scales as $c^{-1/2}$ which corresponds to a correlation length that scales as the Debye length. The crossover density depends on the particular polyelectrolyte.

In Figure 2, we show our calculated q_m for $N_b = 32$. Our data exhibits the same two scaling regimes seen in experiments. The crossover density is not well resolved, but it occurs at about $\rho_m^{(q)} = 0.01\sigma^{-3}$. The overlap density, $\rho_m^* \equiv 1/(\pi\langle R^2\rangle^{3/2}/6)$ is about $0.05\sigma^{-3}$. As will be shown below, the chain begins to significantly contract above $\rho_m^{(q)}$. Beyond this density the intrachain peak is no longer solely dependent on the average chain separation distance. The contraction of the chain is also important. The results on the osmotic pressure and the structure factor peak position show that our simulations model the experimental systems quite well.

Discussion

We can use our calculated structure factors to examine the question of chain length dependence of the individual chain. The form factor is directly related to the individual chain structure. We define the chain form factor as

$$P(\mathbf{q}) = \frac{1}{N_b}|\sum_{j=1}^{N_b}\exp(i\mathbf{q}\cdot\mathbf{r}_j)|^2, \tag{4}$$

where the normalization is $P(0) = N_b$, and we calculate the spherically averaged quantity, $P(q)$. Figure 3 shows $P(q)$ for $N_b = 16, 32, 64$ and 128 at $\rho_m = 5\cdot 10^{-7}\sigma^{-3}$ which is well below the overlap density for all the chain lengths. The plot exhibits the chain length *independence* that holds for all densities. Namely, for $q \gtrsim q_{min} \equiv 2\pi/\langle R^2\rangle^{1/2}$, no chain length dependence exists. This result is a very important since our chains are short in comparison to experiment. Using it we can describe the structure of a chain with $N_b > 64$ on length scales shorter than $\langle R^2(N_b = 64)\rangle^{1/2}$ using the $N_b = 64$ simulations.

Before presenting quantitative results, we graphically display the single chain structure. In Figure 4 (a) and (b) we show typical conformations for $N_b = 32$ at $\rho_m = 1\cdot 10^{-4}\sigma^{-3}$ and $0.3\sigma^{-3}$, respectively. Configuration (a) is at a very dilute density which is three orders of magnitude below the overlap density, and (b) is well above the overlap density. The configurations are oriented by plotting in the plane of the eigenvectors of the inertia tensor with the two largest eigenvalues. The figure clearly shows a transition from stretched to coiled with increasing density. The dilute example is quite stretched but not rodlike. The structure exhibits kinks at short length scales and is bent at long length scales. From such pictures we know that horseshoe shaped fluctuations exist even for longer chains and lower densities. Such structure is not expected at those dilute densities which a persistence length calculation predicts should be fully stretched. The dense case exhibits the coiling reminiscent of neutral good solvent chains.

A convenient quantity to see the transition from stretched to coiled conformation is the ratio of the end-to-end distance and the radius of gyration, $r \equiv \langle R^2\rangle/\langle S^2\rangle$. For ideal chains, $r = 6$; for good solvent chains, $r = 6.3$; for rigid rods, $r = 12$. The plot of r versus ρ_m in Figure 5 shows that the transition from stretched to coiled is present. At the highest densities r approaches the ideal chain value. With decreasing density r rises independent of chain length. In the dilute limit the rigid rod value is not reached for any of the chain lengths. Instead, there is a chain length dependent saturation. This agrees with the expectation that longer chains which have more charges and experience a stronger Coulomb repulsion are straighter. For charged or neutral polymers, the chain ends are always more flexible than the centers.

The saturation of r occurs at $\rho_m \simeq 1\cdot 10^{-4}\sigma^{-3}$, $2\cdot 10^{-5}\sigma^{-3}$ and $1\cdot 10^{-6}\sigma^{-3}$ for $N_b = 16, 32$, and 64 respectively. One might expect this saturation to occur at c_1^*, where chain stiffness is maximal. For these chain length, c_1^* is not defined as the condition $\kappa L > 1$ is not satisfied. However, for $N_b = 128$ using the full Odijk form (14) $c_1^* = 2.8\cdot$

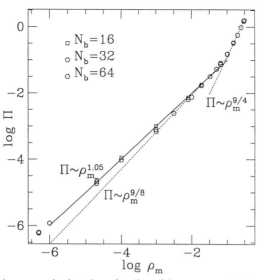

Figure 1. The osmotic pressure is plotted as a function of the monomer density on a log-log scale for various chain lengths. The 16, 32 and 64 bead chains are represented by squares, pentagons and hexagons, respectively. For low densities our relatively small chains tend to have values above the Odijk prediction. This is consistent with the molecular weight dependence found in experiments. At high densities the exponent is 9/4 which is that of neutral chains. (Reproduced with permission from ref. 22. Copyright 1993 APS.)

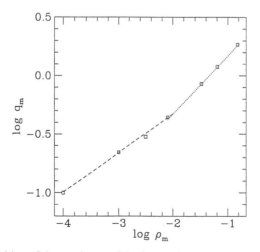

Figure 2. The position of the maximum of the inter-polymer structure factor, q_m, is plotted as a function of monomer density. The same scaling dependence as found in experiments is exhibited. At low densities q_m scales as $\rho_m^{1/3}$ (dashed line) and at high densities q_m scales as $\rho_m^{1/2}$ (dotted line).

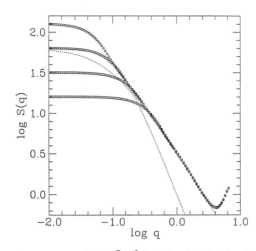

Figure 3. The form factor at $\rho_m = 5 \cdot 10^{-7}\sigma^{-3}$ for $N_b = 16, 32, 64$ and 128. The form factors are chain length independent for $q > 2\pi/\langle R^2 \rangle^{1/2}$. This condition holds for all densities. We are thus able to determine the structure of very long chains on length scales up to the end-to-end distance of our longest chains.

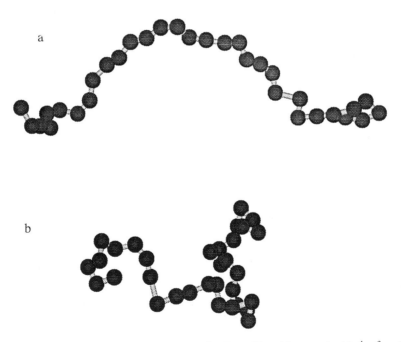

Figure 4. Two typical polymer configurations for $N_b = 32$ at (a) $\rho_m = 1 \cdot 10^{-4}\sigma^{-3}$ and (b) $\rho_m = 0.3\sigma^{-3}$ are shown. At dilute densities the chains are quite stretched (a) and at high densities the chains are coiled (b) like neutral chains.

$10^{-5}\sigma^{-3}$. This value is much too high. One reasonable value of the saturation density, $\rho_m^{(s)}$, is the density at which there is one counterion per polymer volume. For densities below this value, the counterion screening of the intrachain interactions is negligible and one expects r to be constant. For $N_b = 16, 32$ and 64, $\rho_m^{(s)} \equiv 1/(\pi\langle R^2\rangle^{3/2}/6) = 2.3 \cdot 10^{-3}\sigma^{-3}, 2.3 \cdot 10^{-4}\sigma^{-3}$ and $2.3 \cdot 10^{-5}\sigma^{-3}$, respectively. Especially the ratios of these values agree much better with our data.

To compare to scaling predictions, we show the end-to-end distance as a function of monomer density for the same three chain lengths as above in Figure 6. At high densities where $L_e \ll L_i$, $R \sim \rho_m^{-3/16}$ according to Odijk's scaling theory. Our results do not clearly show any scaling regime. However, the solid lines represent the $\rho_m^{-3/16}$ dependence and show that the -3/16 exponent is in agreement with our data. This agreement occurs even though the above criterion is not valid for our simulations. deGennes *et al.* (*10*) have predicted an exponent of -1/8 which is close to Odijk's value and which also fits our data, though not as well as Odijk's value. At lower densities, our results disagree with Odijk's predictions. When $L_e \gg L_i$ at low densities, Odijk predicts $R \sim \rho_m^{-5/16}$, but our curves are flatter than at high densities instead of steeper as Odijk predicts.

Scaling predictions for $P(q)$ have been based on the picture that the polyelectrolyte's structure is composed of straight segments which together form an ideal chain (*1,10,18*). For wavevectors greater than the inverse rodlength, or persistence length, L_p, the form factor should scale as $1/q$. In terms of the exponent ν of the general form, $P(q) \sim 1/q^{1/\nu}$, $\nu = 1$ for the rod. For $q < 2\pi/L_p$, the ideal chain would yield $\nu = 1/2$. However, we have good solvent chains and should see $\nu = 3/5$ at dilute densities.

We show our calculated $P(q)$ spanning dilute and semidilute ranges for the 32 bead chains in Figure 7. Two q-regimes are evident in $P(q)$. A *density and chain length* independent regime occurs at relatively high wavevector ($1 < q\sigma < 2\pi/2$) which corresponds to short length scales up to six bond lengths. At low wavevectors the form factor changes continuously as a function of the monomer density.

The high q-regime dependence is not as expected from present theory. The slope of form factor here is *not* -1. From examination of $q^{1/\nu}P(q)$, we find that for $1 < q\sigma < 2\pi/3$, $\nu = 0.80$ gives the best scaling. This corresponds to chains which are stretched beyond the neutral good solvent value, but well below the rod value. On short length scales, one expects thermal fluctuations to yield values of ν less than one. However, the $\nu = 0.80$ scaling extends to a rather long $6\langle b\rangle$. Furthermore, the chain structure does not become more stretched with increasing length as expected (*10*). This chain length independence is quite surprising, because it implies that the Coulomb interaction can stretch a chain only to a limited extent at least on short length scales. The density independence suggests the existence of a fundamental structural building block of length $6\langle b\rangle$.

The lines in Figure 7 show that at low q the slope varies from about -1 to beyond -5/3, but not quite to -2. Presumably, at higher densities a slope of -2 would be reached. The variation in ν contrasts with the picture of just two values of ν. Evidently, not only the length over which the Coulomb interaction stretches the chain, but also the degree of stretching depends on the screening. At the lowest densities where the form factor has saturated, the value of ν is 0.9. Thus, the chain at very dilute concentrations is stretched more at long length scales than at short length scales, but the chain is not rodlike. The value of ν is the same on short and long length scales at the the overlap density. In the semidilute regime the chain is more contracted at long length scales than at short length scales.

Conclusions

Our results imply a new picture of the chain structure (Figure 8). On short length scales the chain is stretched, but not rodlike. Figure 8 shows tubelets of length $6\langle b\rangle$ which combine to form the complete chain. The tubes have been drawn for simplicity as cylinders. The cylinder radius gives some idea of the curvature of the enveloped monomers. The

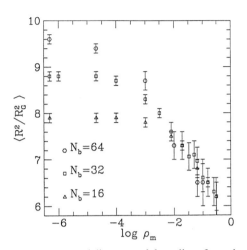

Figure 5. The ratio, r, of the end-to-end distance and the radius of gyration is plotted versus the log of the monomer density. A value of 12 corresponds to a rigid rod and a value of 6 corresponds to a gaussian coil. (Reproduced with permission from ref. 22. Copyright 1993 APS.)

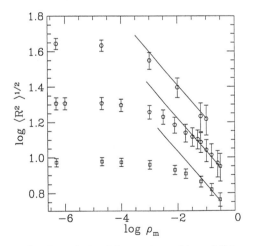

Figure 6. The concentration dependence of the average end-to-end distance agrees with Odijk's scaling prediction which gives an exponent of -3/16. The solid lines show this dependence. The prediction of -1/8 by deGennes *et al.* also is consistent with the data. At low concentrations the curve becomes flatter instead of steeper as predicted by Odijk.

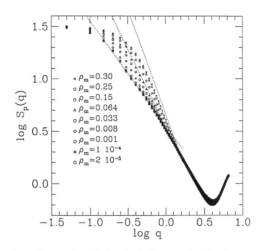

Figure 7. The form factor for 32 bead chains gradually changes from apparent rodlike to coiled form with increasing monomer density. This contradicts the theoretically expected form which is a combination of rigid rod, good solvent chain and ideal chain forms. The dotted lines give slopes of -1, -5/3 and -2 which correspond to the q-dependence of a rigid rod, good solvent chain and ideal chain, respectively. For $0 < \log q\sigma < 0.2$ the slope is steeper than -1, implying the chains are not fully stretched in this regime.

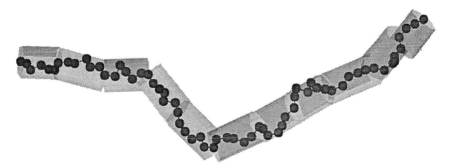

Figure 8. A schematic of our picture of polyelectrolyte structure is shown for a 64 bead chain. The chain is wrapped by a tube composed of tubelets made of length $6\langle b \rangle$. For this length scale, the polymer structure is concentration independent. On longer length scales the chain structure is density dependent.

chain of these tubelets depends on the density or, equivalently, on the degree of screening. At low densities the chain of tubelets is highly stretched but not rodlike. As the density increases the chain become gradually more coiled on long length scales, but the tubelet structure is unchanged. In the semidilute regime the structure is similar to a neutral chains composed of the tubelets. When the strand-strand distance becomes shorter than the tubelet length, the tubelet structure does change.

We would like to acknowledge fruitful discussions with P.A. Pincus and a large grant of computer time from the Höchstleistungsrechenzentrum, Germany within the Disordered Polymers Project. A NATO Travel Grant is greatfully acknowledged.

Literature Cited

(1) Hayter, J.; Janninck, G.; Brochard-Wyart, F.; and de Gennes, P.; *J. Physique Lett.* **1980**, *41*, 451.

(2) Wang, L.; and Bloomfield, V.; *Macromolecules* **1990**, *23*, 804.

(3) Witten, T.; and Pincus, P.; *Europhys. Lett.* **1987**, *3*, 315.

(4) de Gennes, P.; *Scaling Concepts in Polymer Physics*; Cornell University Press: Ithaca, New York, 1979.

(5) Kim, M.; and Pfeiffer, D.; *Euro. Phys. Lett.* **1988**, *5*, 321.

(6) Christos G. A.; Carnie, S. L. *J. Chem. Phys.* **1989**, *91*, 439. Reed, C.; Reed, W *J. Chem. Phys.* **1990**, *92*, 6916. Brender, C; Danino, M *J. Chem. Phys.* **1992**, *97*, 2119. Barrat, J. L.; Boyer, D. *J. Phys. II France* **1993**, *3*, 343. See also references therein.

(7) Kremer, K.; and Grest, G.; *J. Chem. Phys.* **1990**, *92*, 5057.

(8) Dünweg, B.; and Kremer, K.; *Phys. Rev. Lett* **1982**, *66*, 2996.

(9) Adams, D.; and Dubey, G.; *J. Comp. Phys.* **1987**, *72*, 156.

(10) de Gennes, P.; Pincus, P.; and Velasco, R.; *J. Physique* **1976**, *37*, 1461.

(11) Kuhn, W.; Kunzle, D.; and Katchalsky, A.; *Helv. Chim. Acta* **1948**, *31*, 1994.

(12) Hermans, J.; and Overbeek, J.; *Recents Travaux Chimiques* **1968**, *67*, 761.

(13) Skolnick, J.; and Fixman, M.; *Macromolecules* **1977**, *10*, 944.

(14) Odijk, T.; *J. Polym. Scil, Polym. Phys. Ed.* **1977**, *15*, 477.

(15) Odijk, T.; *Polymer* **1978**, *19*, 989.

(16) Fixman, M.; *J. Chem. Phys.* **1982**, *76*, 6346.

(17) Bret, M. L.; *J. Chem. Phys.* **1982**, *76*, 6242.

(18) Odijk, T.; *Macromolecules* **1979**, *12*, 688.

(19) Cates, M.; *J.Phys. II France* **1992**, *2*, 1109.

(20) Wang, L.; and Bloomfield, V.; *Macromolecules* **1991**, *24*, 5791.

(21) Stevens, M.; and Kremer, K.; *Macromolecules* **1993**, *26*, 4717.

(22) Stevens, M.; and Kremer, K.; to be published in Phys. Rev. Lett.

RECEIVED August 6, 1993

Chapter 6

Electrophoresis of Nonuniformly Charged Chains

John L. Anderson and Yuri Solomentsev

Department of Chemical Engineering, Carnegie Mellon University, Pittsburgh, PA 15213

A hydrodynamic analysis of the electrophoretic motion of chain-like particles is presented. The key feature is that the charge is nonuniformly distributed on the particle. Three particle geometries are considered: (1) a polymer-colloid complex, which might arise when a polyelectrolyte attaches itself to several colloidal particles; (2) straight filaments, which could be a model for the coagulation of colloidal spheres or the bundling of charged fibers; and (3) a toroidal ring, which might be a model for DNA plasmids. The theory is based on balancing hydrodynamic and electrophoretic driving forces that operate unevenly on different parts of the heterogeneous particle. The zeta potential, defined as the ion potential at the plane of shear between the suspending liquid (water) and the particle, is allowed to be a function of position along the chain. Although the theory is more general, here we consider for simplicity the case where the Debye screening length is smaller than the radius of the chain. Two interesting results emerge. First, the mobility of a particle is not proportional to its mean charge (or, zeta potential), and hence neutral particles can have significant electrophoretic mobilities given certain distributions of the charge. The second interesting result is that the electrophoretic mobility of a chain of spheres of equal size but different charge depends on the relative placement of the spheres. An example is presented to show that a chain of spheres that is neutral overall could have a zero, positive or negative mobility depending on the relative positions of the spheres of different charge.

Electrophoresis is used to characterize, transport and separate charged macromolecules and colloidal particles in electric fields. The electrophoretic mobility, defined as the velocity of charged particles per unit electric field, is related to the charged state of the particles (1). Two extreme cases are defined in terms of the size of the particle (a) compared to the Debye screening length of the solution (κ^{-1}). If $a \ll \kappa^{-1}$ then the particle looks like a point charge and the counterion cloud has a negligible screening effect; for a sphere of charge q the electrophoretic velocity is given by

0097–6156/94/0548–0067$06.00/0

$$U = \frac{q}{6\pi\eta a}\, \mathbf{E}_\infty \tag{1}$$

where η is the coefficient of viscosity of the fluid and $\mathbf{E}_\infty$ is the applied electric field.

At the other extreme, $a \gg \kappa^{-1}$, the charge on the surface is highly screened; Smoluchowski's equation gives the electrophoretic velocity:

$$U = \frac{D\zeta}{4\pi\eta}\, \mathbf{E}_\infty \tag{2}$$

where ζ is the "zeta potential" of the particle's surface, which is taken to be the plane of shear between the particle and the surrounding liquid. Note that electrostatic units of charge (e.s.u.) are used here, and D is the dielectric constant of the liquid (≈ 78 for water). The classical Gouy-Chapman theory of the electrical double layer gives the following relation between charge/area on the surface (σ) versus zeta potential:

$$\sigma = \frac{DkT}{2\pi z e}\, \kappa \sinh\left(\frac{ze\zeta}{2kT}\right) \approx \frac{D\kappa\zeta}{4\pi} \tag{3}$$

where the approximation holds when $|ze\zeta/kT|$ is order unity or less. (z is the valence of the ions of the electrolyte.) Combining equations 2 and 3 we have the following for the highly screened case when $|\zeta| \leq kT/ze$:

$$U = \frac{q}{6\pi\eta a}\, \frac{3}{2}\, \frac{1}{\kappa a}\, \mathbf{E}_\infty \tag{4}$$

where q is the total charge on the particle ($q = 4\pi a^2 \sigma$).

Equation 2 applies to particles of any shape as long as (i) ζ is uniform over the surface of the particle, and (ii) $(\kappa a)^{-1} \exp|(ze\zeta/2kT)| \ll 1$ where "a" is the characteristic size of the particle (e.g., radius of a sphere). General results are available for arbitrary κa and ζ for uniformly charged spheres (2,3,4). For particles satisfying constraint (ii), models are available to account for arbitrary distributions of ζ on the surface of ellipsoids (5). These particles can translate and *rotate* depending on the distribution of charge; Smoluchowski's equation predicts no rotation. A theory is also available for chains composed of two (6) and three (7) spheres each differing in zeta potential. Experiments (8) with colloidal doublets are consistent with the theory.

In this paper we are concerned with chain-like particles for which the charge (or zeta potential) varies along the chain. Three basic geometries are considered, as shown in Figure 1. The next section deals with a model for the polymer-colloid complex. When hydrodynamic interactions among the various subunits are neglected, an analytic expression for the electrophoretic velocity of the complex is obtained. The analysis of electrophoresis of straight chains is based on the hydrodynamic theory of "slender bodies". Results for long prolate spheroids and right circular cylinders are presented. The results for the circular cylinder are used to model a straight chain of equal spheres attached to each other. Finally, the hydrodynamic analysis by Johnson and Wu (9) is used to model the electrophoresis

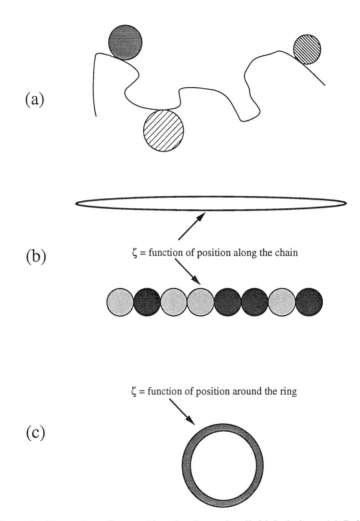

Figure 1. Examples of nonuniformly charged colloidal chains. (a) Polymer-colloid complex; (b) straight chains having a slender shape; (c) toroidal ring. In all cases the charge density varies with position along the chain.

of a toroidal ring having a variable charge along its contour. For all three geometries shown in Figure 1, our theory predicts that neutral chains will move in electric fields if the distribution of the charge along the chain has certain properties.

POLYMER-COLLOID COMPLEXES

Consider N spherical colloids connected through their attachment to a polymer chain, as shown in Figure 1a. If particle "i" were not constrained by its attachment to the polymer, then its velocity in the applied electric field (E_∞) would be simply $\mu_i E_\infty$ where μ_i is the electrophoretic mobility; for the highly screened case the mobility is given by the coefficient of E_∞ in equation 2 or 4. Likewise, the polymer would move at velocity $\mu^{(p)} E_\infty$ in the absence of the attached colloids. If the complex is a rigid body, then the polymer and attached colloids must all move at the same velocity U in the field. The total hydrodynamic force is found by summing up the friction on each element, which is proportional to the difference between U and the unconstrained electrophoretic velocity of the element:

$$F = -\sum_{i=1}^{N} f_i \left[U - \mu_i E_\infty \right] - f^{(p)} \left[U - \mu^{(p)} E_\infty \right] = 0 \tag{5}$$

The f_i and $f^{(p)}$ are orientation-averaged friction coefficients of the colloids and the polymer, respectively. Rotation of the complex is neglected; this is justified if the ensemble-averaged angular velocity is zero, either because of strong Brownian motion or by alignment of the dipole of the complex with the electric field. If dipole alignment is a factor, the electrophoretic mobility (U/E_∞) would be a function of the applied field.

$\quad$ F must equal zero because in the screened case the electric field exerts no force on the elements if their double layers are included *(10)*. Thus, the velocity of the complex is given by

$$U = \frac{f^{(p)} \mu^{(p)} + \sum_{i=1}^{N} f_i \mu_i}{f^{(p)} + \sum_{i=1}^{N} f_i} E_\infty \tag{6}$$

If we neglect all hydrodynamic interactions in the complex and consider only spherical colloids, then $f_i = 6\pi\eta a_i$. If we further assume that the polymer's structure is not significantly perturbed by the attached colloids, we have $f^{(p)} = 6\pi\eta R_H$ where R_H is the Stokes-Einstein (hydrodynamic) radius which is calculated from the diffusion coefficient of the polymer. Combining these expressions into equation 6 produces

$$U = \frac{R_H \mu^{(p)} + \sum_{i=1}^{N} a_i \mu_i}{R_H + \sum_{i=1}^{N} a_i} E_\infty \tag{7}$$

The above is the "free draining" result for the complex because colloid-colloid and colloid-polymer hydrodynamic interactions are neglected; however, hydrodynamic interactions among elements of the polymer chain are included in $f^{(p)}$, or equivalently R_H. Because hydrodynamic and electrostatic interactions among the colloids are neglected, the velocity of the complex is independent of the relative spacing and orientation of the colloids on the polymer chain.

As an example, assume all N colloids are identical with radius "a" and electrophoretic mobility in free solution equal to μ_o. The dimensionless electrophoretic mobility of the complex, $U/\mu_o E_\infty$, is given by

$$\frac{U}{\mu_o E_\infty} = \frac{\frac{\mu^{(p)}}{\mu_o} + \left(\frac{a}{R_H}\right) N}{1 + \left(\frac{a}{R_H}\right) N} \tag{8}$$

This expression is plotted in Figure 2 for the case $a/R_H = 0.5$. (Note that a/R_H equals the ratio of polymer-to-colloid diffusion coefficients, each of which is a measurable property before the complex is formed.) For the case $\mu^{(p)} = \mu_o$ the velocity of the complex is independent of N and $U = \mu_o E_\infty$; this is expected because all the elements of the complex (N colloids plus the polymer chain) move at $\mu_o E_\infty$. For the case where $\mu^{(p)}$ and μ_o have opposite signs, the direction of electrophoresis changes as N increases.

Experiments have been performed with silica particles bound to polyelectrolyte chains *(11)*. One objective of such measurements is to relate the number of adsorbed colloids (N) to the electrophoretic mobility of the complexes. Equation 8, though approximate because of the free-draining model, indicates that N can be deduced from the electrophoretic mobility if the mobilities of the polyelectrolyte ($\mu^{(p)}$) and micelles (μ_o) are known *a priori*.

Another interesting question is how the mobility of the complex is related to the total charge. To investigate this using the free-draining model, we must assume a model relating the mobility of each subunit to the charge. If we assume modest zeta potentials ($|\zeta|$ comparable to kT/ze) and highly screened double layers for the colloids ($\kappa a \gg 1$), we have the following from equation 4:

$$\mu_o = \frac{q_o}{4\pi\eta\kappa a^2} \tag{9}$$

where q_o is the charge on each colloid. The relationship between mobility and charge for the polymer chain is more ambiguous, depending on the configurational model for the chain *(12,13)*. Here, for illustration purposes only, we consider a general linear model:

$$\mu^{(p)} = W \frac{q^{(p)}}{2\pi\eta l \, \kappa b} \tag{10}$$

where b is the radius and l the length of the chain. The coefficient W is a function of κb and the contour of the chain; here we take it to be unity, which would be the case for a highly screened ($\kappa b \gg 1$), extended chain. In Figure 3 we re-plot the calculations for $\mu^{(p)}/\mu_0 = -2$ shown in Figure 2. These graphs show that neutral complexes will move and zero-mobility does not imply charge neutralization. This result would not significantly change had we chosen a different value for W.

Another example of the motion of neutral complexes is given by two colloids of equal but opposite charge ($q_1 = -q_2$) connected by a neutral polymer ($\mu^{(p)} = 0$) having a negligible hydrodynamic resistance ($R_H \ll a_1$ or a_2). For cases when the mobility of the colloids is given by equation 4, the velocity of the complex is

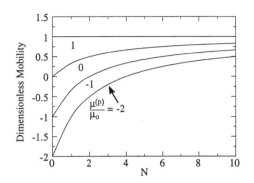

Figure 2. Dimensionless mobility of a polymer-colloid complex (mobility of the complex, U/E_∞, divided by the mobility of the colloids, μ_0) as a function of the number of colloids attached to the polymer. The curves are from equation 8 with $a/R_H = 0.5$.

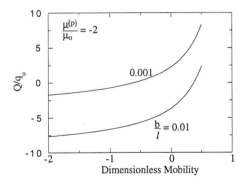

Figure 3. The total charge (Q) of the complex versus the dimensionless mobility (defined in Figure 2) of a polymer-colloid complex. q_0 is the charge on each colloid attached to the polymer. The curves were computed from equations 8-10 with $a/R_H = 0.5$ and $\mu^{(p)}/\mu_0 = -2$. The hydrodynamic radius of the polymer chain was determined from the theory for a rod of length l and radius b: $R_H = l/[3 \ln (l/b)]$.

$$U = \frac{q_1}{\varepsilon \kappa a_1^2} \left[\frac{1 - \dfrac{a_1}{a_2}}{1 + \dfrac{a_2}{a_1}} \right] E_\infty \tag{11}$$

Clearly the complex moves in the field if the radii of the two colloids differ, even though the *net charge on the entire complex is zero*. An interesting prediction of equation 11 is that the direction of electrophoresis is determined by the *smaller* colloid. Hydrodynamic analysis of nonuniformly charged spheroidal particles also shows that such particles can have a significant electrophoretic mobility even when they are neutral *(5)*.

A more quantitatively accurate model for complexes would allow for hydrodynamic interactions among the colloidal subunits. If the hydrodynamic effects of the connecting polymer chain are negligible ($R_H \ll a$), the force on each colloidal subunit can be written in terms of general friction tensor (ξ_i) which is a function of the size (a_i) and relative position ($\{r_j\}$, $j = 1 \rightarrow N$) of all the spherical particles in the complex:

$$F_i = -\xi_i \cdot [U - \mu_i E_\infty] \tag{12}$$

where $\xi(\{r_j\})$ accounts for hydrodynamic interactions among the particles; in the absence of these interactions ξ_i equals $(6\pi\eta a_i)I$ where I is the unit tensor. Calculations based on the Oseen approximation for hydrodynamic interactions have been used to model the hydrodynamic resistance of particles of different geometry *(14,15)*. The velocity of the complex is found by equating the sum all the forces on the N subunits to zero:

$$U = \left(\sum \xi_i\right)^{-1} \cdot \sum \mu_i \, \xi_i \cdot E_\infty \tag{13}$$

This formula is valid if rotational effects are negligible, that is, $\langle \Omega \rangle = 0$. Because the velocity is a function of the relative configuration of the particles, $\{r_j\}$, the measurable velocity is the ensemble average of equation 13. Note that if all the subunits have the same electrophoretic mobility, $\mu_i = \mu_o$, then the velocity of the complex is simply $\mu_o E_\infty$ no matter what the relative configuration of the subunits.

Equations 12 and 13 do not allow for interactions among the subunits in the electrophoresis terms; that is, μ_i is assumed independent of $\{r_j\}$. In principle the electrophoretic mobilities μ_i are also tensors that are functions of the relative configuration of the particles. However, interactions among spheres undergoing electrophoresis are extremely weak *(10,16)* and neglect of these interactions is probably justified given the simplicity of the model itself.

ELECTROPHORESIS OF SLENDER PARTICLES

Figure 4 shows the general model of a "slender particle" which may possess a distribution of charge along its contour. The contour length equals 2L and "s" is the dimensionless position measured along the contour ($s = -1$ and $+1$ at the ends). The cross section at any contour position is a circle of radius b which varies with s, and $\lambda(s)$ is defined as $b(s)/b_{max}$. The three length scales of the system have the

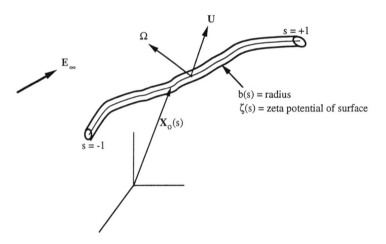

Figure 4. Geometry of a slender particle. L is the half-length of the contour, and s is the distance along the contour from the center divided by L. The cross-section of the particle is a circle whose radius (b) varies with s. The contour of the particle is defined by the function $\mathbf{X}_0(s)$. The particle is assumed to be a rigid body with the velocity of the center equal to $\mathbf{U}$ and the angular velocity equal to Ω.

following scaling: $\kappa^{-1} \lesssim b_{max} \ll L$. The slenderness parameter (ε) is defined as b_{max}/L. The locus of points along the axis of the contour is defined by the vector function $\mathbf{X}_0(s)$, and the position vector is defined as $\mathbf{r} = \mathbf{X}_0(s) - \mathbf{X}_0(0)$. The contour can be curved as long as its radius of curvature is much greater than b_{max}.

A continuous representation of the "subunit model" introduced in the previous section is the basis of the model for slender particles. A subunit of dimensionless length ds would have an electrophoretic velocity equal to $\mu(s)\cdot\mathbf{E}(s)$ where $\mathbf{E}$ is the local electric field and the tensor μ depends on the zeta potential at s. Because the particle is locally a cylinder, μ has two principle values, one for the field parallel to the axis (i.e., parallel to the local tangent of $\mathbf{X}_0(s)$) and the other for the field perpendicular to the axis *(17)*. If $\kappa^{-1} \ll b(s)$ then $\mu(s) = (D/4\pi\eta)\ \zeta(s)\ \mathbf{I}$; we make this assumption here.

The slender particle is assumed to be a rigid body. $\mathbf{U}$ is the translational velocity of the center (s = 0) and Ω is the angular velocity. The force per unit length, $\Gamma(s)$, is given by

$$\Gamma(s) = -\xi(s) \bullet [\mathbf{U} + \Omega \times \mathbf{r}(s) - \mu(s)\mathbf{E}] \qquad (14)$$

where the hydrodynamic resistance tensor, $\xi(s)$, allows for hydrodynamic interactions along the contour of the particle. The determination of the local electric field involves solution of Laplace's equation for the electrical potential outside the double layer. Cole *(18)* showed that to $O(\varepsilon^2 \ln\varepsilon)$, $\mathbf{E}$ equals the applied field ($\mathbf{E}_\infty$) which is a constant. $\mathbf{U}$ and Ω are determined by setting the total force and torque on the particle equal to zero:

$$\int_{-1}^{+1} \Gamma(s)ds = 0 \qquad (15a)$$

$$\int_{-1}^{+1} \mathbf{r}(s) \times \Gamma(s)ds = 0 \qquad (15b)$$

The remaining problem, then, is to determine $\xi(s)$ for translation and rotation given the geometry of the particle, $\mathbf{X}_0(s)$ and $\lambda(s)$. Note that if the particle has certain symmetry, for example is straight and symmetric ($\lambda(s) = \lambda(-s)$), then translation and rotation are uncoupled; in this case equation 15a yields $\mathbf{U}$ and equation 15b yields Ω.

Theories for the hydrodynamics of slender particles are based on determining the distribution of "Stokeslets" along the contour of the chain. The Stokeslet distribution $\alpha(s)$ is defined such that for a given motion of the particle the force per unit length exerted by the fluid on the particle is given by $-8\pi\eta\alpha$. In terms of the resistance tensor ξ we have

$$\alpha(s) = \frac{1}{8\pi\eta}\ \xi(s) \bullet \mathbf{V}(s) \qquad (16)$$

where $\mathbf{V}(s)$ is the velocity of position s of the particle. From equation 14 we have

$$\mathbf{V}(s) = \mathbf{U} + \Omega \times \mathbf{r}(s) - \mu(s)\mathbf{E}_\infty \qquad (17)$$

The velocity of the particle is determined by substituting equations 16 and 17 into equation 15. The Stokeslet functions for translation $(\mathbf{U})$ and rotation$(\mathbf{\Omega})$ are evaluated using the theory of Cox (19), which is correct to $O(1/\ln \varepsilon)^2$. By using the Lorentz reciprocal theorem (5) we avoid the calculation of the Stokeslet distribution for electroosmotic flow $(\mathbf{V} = -\mu(s)\mathbf{E}_\infty)$. Below are some results for the shapes shown in Figure 1. Details of the hydrodynamic analysis are presented elsewhere (20).

LONG SPHEROID $\left(\lambda(s) = \sqrt{1-s^2}\,\right)$

For this particular shape the Stokeslet distribution can be determined to an error of $O(\varepsilon^2\ln\varepsilon)$. The three Stokeslet functions are:

$$\alpha_\| = \frac{1}{2\,[2\ln(2/\varepsilon) - 1]} \tag{18a}$$

$$\alpha_\perp = \frac{1}{2\ln(2/\varepsilon) + 1} \tag{18b}$$

$$\alpha_r = \frac{s}{2\ln(2/\varepsilon) - 1} \tag{18c}$$

where $\alpha_\|$ and $\alpha_\perp$ are for translation at unit velocity parallel and perpendicular to the axis of rotational symmetry of the spheroid, respectively, and α_r is the Stokelet for rotation perpendicular to the axis. Equations 14-16 then give the following for any distribution of mobilities along the chain:

$$\mu_\| \equiv \frac{U_\|}{E_\|} = <\mu> \tag{19a}$$

$$\mu_\perp \equiv \frac{U_\perp}{E_\perp} = <\mu> \tag{19b}$$

$$\mu_r \equiv \frac{\Omega L}{E_\perp} = 3 <s\mu> \tag{19c}$$

where the brackets define an average over the contour length:

$$<f> = \frac{1}{2} \int_{-1}^{+1} f \, ds \tag{20}$$

For the highly screened case, $\kappa b(s) \gg 1$, we can substitute the following for the local mobility:

$$\mu(s) = \frac{D}{4\pi\eta} \zeta(s) \tag{21}$$

For zeta potentials of order kT/e or less, the charge per unit length of the particle is proportional to $\zeta(s)$. At first glance it might appear that the mobilities in equations 19 are proportional to the average potential, or charge, on the chain, but this is not the case. If we let ζ_{ave} denote the average of ζ over the surface area of the particle, then the translational mobilities in equations 19 are re-expressed as

$$\mu_{\parallel} = \mu_{\perp} = \frac{D}{4\pi\eta} \left\{ \zeta_{ave} + \frac{2}{\pi} \int_{-1}^{+1} \left[\frac{\pi}{4} - (1-s^2)^{1/2} \right] \zeta(s) ds \right\} \qquad (22)$$

Thus, for distributions of ζ that depend on even powers of s, the electrophoretic mobility of the chain is not zero even if ζ_{ave} equals zero; that is, neutral chains will move in an electric field if the distribution of the zeta potential is an even function of s. For the distribution $\zeta = \zeta_0 + \zeta_1 s + \zeta_2(3s^2-1)$, where the ζ_i are constant, equation 22 yields

$$\mu_{\parallel} = \mu_{\perp} = \frac{D}{4\pi\eta} \left[\zeta_{ave} + \frac{1}{4} \zeta_2 \right] \qquad (23a)$$

$$\mu_r = \frac{D}{4\pi\eta} \zeta_1 \qquad (23b)$$

STRAIGHT CIRCULAR CYLINDER ($\lambda(s) = 1$)

Using Cox's model *(19)* we obtain the following Stokeslet distributions:

$$\alpha_{\parallel} = \frac{1 - 2 \ln\left(2\varepsilon \sqrt{1 - s^2}\right)}{8 \ln^2 \varepsilon} \qquad (24a)$$

$$\alpha_{\perp} = - \frac{1 + 2 \ln\left(2\varepsilon \sqrt{1 - s^2}\right)}{4 \ln^2 \varepsilon} \qquad (24b)$$

$$\alpha_r = \frac{-s\left(1 + 2 \ln\left(2\varepsilon \sqrt{1 - s^2}\right)\right)}{4 \ln^2 \varepsilon} \qquad (24c)$$

The largest errors in these expressions occur at the ends s = ± 1. Because these results have an error of order $(1/\ln\varepsilon)^3$ they are probably only valid for $\varepsilon \leq 0.1$. Using equations 14-16 we have

$$\mu_{\parallel} = \frac{D}{4\pi\eta} \frac{\left\langle \zeta(s) \left[1 - 2 \ln\left(2\varepsilon \sqrt{1-s^2}\right) \right] \right\rangle}{3 - 2 \ln(4\varepsilon)} \qquad (25a)$$

$$\mu_\perp = -\frac{D}{4\pi\eta} \frac{\langle \zeta(s)[1 + 2\ln(2\varepsilon\sqrt{1-s^2})]\rangle}{1 - 2\ln(4\varepsilon)} \tag{25b}$$

$$\mu_r = -9\frac{D}{4\pi\eta} \frac{\langle s\,\zeta(s)[1 + 2\ln(2\varepsilon\sqrt{1-s^2})]\rangle}{5 - 6\ln(4\varepsilon)} \tag{25c}$$

For a cylinder the area-average zeta potential (ζ_{ave}) equals $\langle\zeta\rangle$, so it is apparent from equation 25 that if $\zeta(s)$ is an even function then the particle will move even if it is neutral. Given the distribution $\zeta = \zeta_0 + \zeta_1 s + \zeta_2(3s^2-1)$, where $\zeta_{ave} = \zeta_0$, equations 25 give

$$\mu_\| = \frac{D}{4\pi\eta}\left[\zeta_0 + \frac{2}{3(3 - 2\ln(4\varepsilon))}\zeta_2\right] \tag{26a}$$

$$\mu_\perp = \frac{D}{4\pi\eta}\left[\zeta_0 + \frac{2}{3(1 - 2\ln(4\varepsilon))}\zeta_2\right] \tag{26b}$$

$$\mu_r = \frac{D}{4\pi\eta}\zeta_1 \tag{26c}$$

The parallel and perpendicular mobilities are different; in fact, the mobility perpendicular to the cylinder's axis is slightly *greater* than for parallel to the axis if ζ_2 and ζ_0 have the same sign.

CHAIN OF EQUAL SIZE SPHERES

The results for a straight cylinder can be used to estimate the electrophoretic mobility of a straight chain of touching spheres all of which are the same size but could have different zeta potentials. For the highly screened case ($\kappa a \gg 1$) we have the general formula

$$\mu = \frac{D}{4\pi\eta}\sum_{i=1}^{N} A_i\zeta_i \tag{27}$$

where ζ_i is the zeta potential of sphere i and N is the number of spheres in the chain. Three possible motions are possible: translation parallel to the chain ($\mu_\| = U_\|/E_\|$), translation perpendicular to the chain ($\mu_\perp = U_\perp/E_\perp$) and rotation about an axis perpendicular to the chain ($\mu_r = Na\Omega/E_\perp$). The chain is approximated by a straight cylinder with $\varepsilon = N^{-1}$; there are N equal sections of potential ζ_i. The mobilities, and hence A_i, are calculated from equations 25 with $\mu = \varepsilon\zeta/4\pi\eta$. Table I gives the formulas for calculating the A_i, and a plot of A_i for N = 10 is made in Figure 5. From symmetry considerations we know the following

$$\mu_\| \text{ and } \mu_\perp: \sum_{i=1}^{N} A_i = 1 \;;\; A_{N+1-i} = A_i \tag{28a}$$

TABLE I.   Mobility of a chain of N particles of the same size but different zeta potential. The coefficients A_i appear in equation 27. $\varepsilon = N^{-1}$ where ε is the diameter/length of the equivalent cylinder

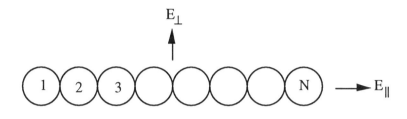

$\mu_{\parallel} = U_{\parallel}/E_{\parallel}$:

$$A_i = \Big\{ 2\left[(1-i\varepsilon)\ln(1-i\varepsilon) - \ln\left[1-(i-1)\,\varepsilon\right]\right.$$
$$\left. - i\varepsilon\ln(4i\varepsilon)\right] + \varepsilon\left[6-4\ln(2\varepsilon) - 2(1-i)\ln\left[4\varepsilon\,(i-1)\right.\right.$$
$$\left.\left.\left(1-(i-1)\,\varepsilon\right)\right]\right]\Big\}\left[6-4\ln(4\varepsilon)\right]^{-1} \tag{1.1}$$

$\mu_{\perp} = U_{\perp}/E_{\perp}$:

$$A_i = \Big\{ 2\left[(1-i\varepsilon)\ln(1-i\varepsilon) - \ln\left[1-(i-1)\,\varepsilon\right]\right.$$
$$\left. - i\varepsilon\ln(4i\varepsilon)\right] + \varepsilon\left[6-4\ln(2\varepsilon) - 2(1-i)\ln\left[4\varepsilon\,(i-1)\right.\right.$$
$$\left.\left.\left(1-(i-1)\,\varepsilon\right)\right] - 4\right]\Big\}\left[2-4\ln(4\varepsilon)\right]^{-1} \tag{1.2}$$

$\mu_r = Na\Omega/E_{\perp}$:

$$A_i = \frac{C}{2}\left[(2\varepsilon i - 1)^2 - (2\varepsilon(i-1) - 1)^2\right]$$
$$+ \frac{D}{4}\Big\{-2\,(\varepsilon i - 1)^2 + (2\varepsilon(i-1) - 1)^2 + \left((2\varepsilon i - 1)^2 - 1\right)$$
$$\ln\left[4\varepsilon i\,(1-\varepsilon i)\right] - \left[(2\varepsilon(i-1) - 1)^2 - 1\right]\ln\left[4\varepsilon(i-1)\,(1-\varepsilon(i-1))\right]\Big\} \tag{1.3}$$

where

$$C = \frac{9}{2}\,\frac{1 + 2\ln(2\varepsilon)}{-5 + 6\ln(4\varepsilon)}$$

$$D = \frac{9}{2}\,\frac{1}{-5 + 6\ln(4\varepsilon)}$$

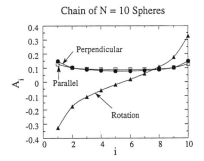

Figure 5. The coefficient A_i versus i for a chain of 10 spheres (see equation 27). The symbols have the following meanings: $\| \to \mu_\|$; $\perp \to \mu_\perp$; $r \to \mu_r$.

$$\mu_r : \sum_{i=1}^{N} A_i = 0 \; ; \qquad A_{N+1-i} = -A_i \tag{28b}$$

The spheres at either end of the chain have the greatest influence on the electrophoretic mobility. As an example of the importance of the location of the charges, consider a chain of 10 spheres only two of which are charged. In Table II the mobility of the chain is shown as a function of the positioning of the two charged spheres for a) the same ζ for both spheres and b) the same $|\zeta|$ for both spheres but the signs are opposite. Note that in case b) the chain is *neutral*.

For an ensemble of chains with a *random* positioning of the spheres, the average mobilities are given by the following:

$$<\mu_{\parallel}> = <\mu_{\perp}> = \frac{D}{4\pi\eta} <\zeta> \tag{29a}$$

$$<\mu_r> = 0 \tag{29b}$$

where $<\zeta>$ is the statistical-mean zeta potential which is the same for all i. This is the result expected from intuition. In example (a) of Table II, $<\zeta> = 0.2 \zeta_0$; in example (b), $<\zeta> = 0$. If, however, the relative positioning of the different types of spheres is *not random*, then equations 29 are invalid.

TORUS ($\lambda(s) = 1$)

A torus is a circular cylinder whose contour is a circle. The Stokeslet distribution for the torus was derived by Johnson and Wu *(9)* to order $\varepsilon^2 \ln\varepsilon$; these functions of s are listed in the Table III. The electrophoretic motion is described by defining two mobilities:

$$\mathbf{U} = \mathbf{M_T} \cdot \mathbf{E}_\infty \; ; \qquad \mathbf{\Omega} = \mathbf{M_R} \cdot \mathbf{E}_\infty \tag{30}$$

We consider the distribution $\zeta = \zeta_0 + \zeta_1 s + \zeta_2 (3s^2-1)$. The two mobilities defined above can be expressed as

$$\mathbf{M_T} = \zeta_0 \mathbf{I} + \zeta_1 \mathbf{M_T^{(1)}} + \zeta_2 \mathbf{M_T^{(2)}} \tag{31a}$$

$$\mathbf{M_R} = \zeta_1 \mathbf{M_R^{(1)}} + \zeta_2 \mathbf{M_R^{(2)}} \tag{31b}$$

Expressions for the dipole ($\sim \zeta_1$) and quadrupole ($\sim \zeta_2$) mobilities are shown in Table IV.

The area-averaged zeta potential of the torus equals ζ_0; thus, for modest potentials the charge on the torus is proportinal to ζ_0. Equations 31 clearly show that even if the torus is *neutral* there is rotation and translation proportional to the dipole and quadrupole moments, respectively. As an example consider $\zeta_0 = \zeta_1 = 0$. A torus with $\varepsilon = 0.1$ will experience the following velocities for different orientations of the electric field.

TABLE II.   Mobility of a chain of 10 spheres only two of which are charged.
The dimensionless mobility is defined as $U_{\parallel}/\mu_0 E_{\parallel}$ where μ_0
($=\varepsilon\zeta_0/4\pi\eta$) is the mobility of sphere ●. $A_1 = 0.1280$, $A_2 = 0.1016$,
$A_3 = 0.0934$, $A_4 = 0.0894$, $A_5 = 0.0876$. $A_{11-i} = A_i$

a) The two charged spheres have the same zeta potential

Configuration	Dimensionless Mobility
$\rightarrow$ E$_{\parallel}$	
●○○○○○○○○●	0.256
○○○○●●○○○○	0.175
●○○○○●○○○○	0.216

b) The two spheres have zeta potentials of equal magnitude but opposite
sign (i.e., the chain is *neutral*)

Configuration	Dimensionless Mobility
$\rightarrow$ E$_{\parallel}$	
●○○○○⊗○○○○	+0.040
⊗○○○○●○○○○	-0.040
●○⊗○○○○○○○	+0.035
⊗○●○○○○○○○	-0.035
●○○○○○○○○⊗	0.000

TABLE III. Torus of contour length 2L and cross-sectional radius b_0. $\varepsilon = b_0/L$ and $\gamma = \pi\varepsilon$. The Stokeslet distributions are from reference *(9)*. The orthonormal set (e_n, e_s, e_b) is defined such that e_s is tangent to the contour at each position s, e_n is in the plane of the ring and e_b points in the direction "2"

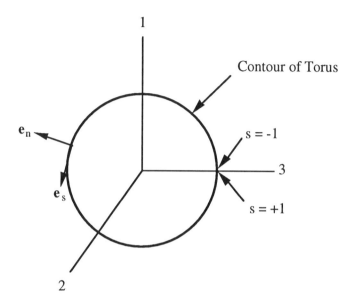

Translation (subscript denotes direction)

$$\alpha_1 = \frac{-3 + \ln(8/\gamma)}{-2 - 5\ln(8/\gamma) + 2\ln^2(8/\gamma)} \sin[\pi(1+s)] \, e_n$$
$$+ \frac{-5 + 2\ln(8/\gamma)}{-8 - 20\ln(8/\gamma) + 8\ln^2(8/\gamma)} \cos[\pi(1+s)] \, e_s$$

$$\alpha_2 = \frac{1}{1 + 2\ln(8/\gamma)} \, e_b$$

Rotation (subscript denotes rotation axis)

$$\alpha_1 = \frac{1}{3 - 2\ln(8/\gamma)} \cos[\pi(1+s)] \, e_b$$

$$\alpha_2 = \frac{1}{-8 + 4\ln(8/\gamma)} \, e_s$$

TABLE IV. Mobility tensors for the torus as defined by equations 31. See Table 3 for the definitions of the Cartesian axes (X_1, X_2, X_3). $\gamma = \pi \varepsilon$. For the tensors $M(i, j)$, j is the direction of the field and i is the direction of the velocity

$$M_T^{(1)}(1,3) = M_T^{(1)}(3,1) = \frac{17 - 6 \ln(8/\gamma)}{2\pi\left[-7 + 2\ln(8/\gamma)\right]}$$

All other $M_T^{(1)}(i,j) = 0$

$$M_T^{(2)}(1,1) = -M_T^{(2)}(3,3) = \frac{3\left[17 - 6\ln(8/\gamma)\right]}{2\pi^2\left[-7 + 2\ln(8/\gamma)\right]}$$

All other $M_T^{(2)}(i,j) = 0$

$$M_R^{(1)}(2,3) = -\frac{1}{2}M_R^{(1)}(3,2) = \frac{1}{\pi}$$

All other $M_R^{(1)}(i,j) = 0$

$$M_R^{(2)}(1,2) = -2M_R^{(2)}(2,1) = -\frac{12}{\pi^2}$$

All other $M_R^{(2)}(i,j) = 0$

Field oriented toward direction "3": $U_3/E_\infty = 0.8\ D\zeta_2/4\pi\eta$

Field oriented toward direction "1": $U_1/E_\infty = -\ 0.8\ D\zeta_2/4\pi\eta,$

$\Omega_2 L/E_\infty = 0.61\ D\zeta_2/4\pi\eta$

Field oriented toward direction "2": $\Omega_1 L/E_\infty = -\ 1.22\ D\zeta_2/4\pi\eta$

This simple example illustrates the interesting electrophoretic dynamics possible when the charge is nonuniformly distributed on a ring.

ACKNOWLEDGMENT

Support for this work came from the National Aeronautics and Space Administration (Grant NANAG8-964) and the Westvaco Corporation.

LITERATURE CITED

1. Hunter, R.J. *Zeta Potential In Colloid Science*; Academic Press: New York, 1981.
2. Derjaguin, B.V.; Dukhin, S.S.   Nonequilibrium Double Layer and Electrokinetic Phenomena. In *Surface and Colloid Science*, Matijevic, E., Ed.; John Wiley & Sons: New York, 1974; Vol. 7, Chapter 3.
3. O'Brien, R.W.; White, L.R. *J. Chem. Soc., Faraday Trans. 2*, **1978**, *74*, 1607.
4. O'Brien, R.W. *J. Colloid Interface Sci.* **1983**, *92*, 204.
5. Fair, M.C.; Anderson, J.L. *J. Colloid Interface Sci.* **1989**, *127*, 388.
6. Fair, M.C.; Anderson, J.L. *Int. J. Multiphase Flow* **1990**, *16*, 663.
7. Keh, H.J.; Yang, F.R. *J. Colloid Interface Sci.* **1991**, *145*, 362.
8. Fair, M.C.; Anderson, J.L. *Langmuir*, **1992**, *8*, 2850.
9. Johnson, R.E.; Wu, T.Y. *J. Fluid Mech.* **1979**, *95*, 263.
10. Anderson, J.L. *Annu. Rev. Fluid Mech.* **1989**, *21*, 61.
11. Cabane, B.; Wong K.; Wang, T.K.; Lafuma, F.; Duplessix, R. *Colloid and Polymer Sci.* **1988**, *266*, 101.
12. Stitger, D. *Phys. Chem.* **1978**, *82*, 1417.
13. Stitger, D. *Phys. Chem.* **1978**, *82*, 1424.
14. Bloomfield, V.; Dalton, W.O.; Van Holde, K.E. *Biopolymers*, **1967**, *5*, 135.
15. De La Torre, J.G.; Bloomfield, V. *Biopolymers* **1978**, *17*, 1605.
16. Keh, H.J.; Chen, S.B. *J. Colloid Interface Sci.* **1989**, *130*, 542, 556.
17. Henry, D.C. *Proc. R. Soc. London*, **1932**, *133*, 106.
18. Cole, J.D. Perturbation Methods in Applied Mathematics. Waltham, MA: Blaisdel, 1968.
19. Cox, R.G. *J. Fluid Mech.* **1970**, *44*, 791.
20. Solomentsev, Y.; Anderson, J.L. "Electrophoresis of Slender Bodies," submitted for publication, **1993**.

RECEIVED September 1, 1993

Chapter 7

Undulation-Enhanced Forces in Hexagonal Gels of Semiflexible Polyelectrolytes

Theo Odijk

Faculty of Chemical Engineering and Materials Science, Delft University of Technology, P.O. Box 5045, 2600 GA Delft, Netherlands

A theoretical analysis is given of undulation-enhanced forces in solutions of semiflexible polyelectrolytes. Undulation enhancement is fairly weak when the system is isotropic, nematic or cholesteric. Electrostatic repulsion can easily be enhanced by an order of magnitude in hexagonal gels. Van der Waals forces are perturbed mildly by configurational fluctuations although their enhancement does affect the free energy function near the secondary minimum. A stability analysis of enhanced electrostatic versus Van der Waals forces is presented in order to discuss experiments on gels of tobacco mosaic virus.

Semiflexible polyelectrolytes are here defined as having a persistence segment of large aspect ratio i.e. the persistence length L_p is much greater than the effective diameter. At high ionic strength, the latter reduces to a geometric diameter D; at low salt, it is approximately proportional to the Debye screening length κ^{-1}. Several important biopolymers are semiflexible polyelectrolytes exhibiting intriguing phase behavior. For instance, aqueous DNA forms a variety of lyotropic liquid crystals. Upon increasing the concentration, the following sequence of phases is seen (*1*): isotropic to cholesteric to two dimensional columnar hexagonal to three dimensional columnar hexagonal and finally to orthorhombic crystalline. There is a possibly optimistic tendency to believe that these phases ought to be understood in fairly simple statistical physical terms. It may be argued that the chemical detail of the polyion backbone is not critically important for long-range electrostatics swamps any other interactions, except possibly those of dispersion type. Furthermore, our qualitative insight in the condensed matter physics of complex systems is increasing each decade, slowly yet steadily.

The purpose of this paper is to show that the electrostatic interaction may be greatly enhanced by undulations or chain fluctuations in the hexagonal phase of positionally ordered polyions. In nonadhesive states, Van der Waals forces are influenced to a much lesser degree at least within the Gaussian approximation adopted here. Nonetheless, undulation enhancement of dispersion forces cannot be neglected when trying to locate the secondary minimum for a gel stabilized by these interactions. Undulation enhancement is of minor import in solutions where positional order is weak or absent as I now show.

0097–6156/94/0548–0086$06.00/0

Isotropic, Nematic and Cholesteric Phases

Undulation enhancement of the electric potential exerted by a polyion was first considered by Odijk and Mandel (2) who developed a perturbation scheme in terms of the large parameter $L_p\kappa$. A similar procedure can be used to evaluate the average interaction between two nearby undulating polyions. However, I here present only a qualitative analysis for wormlike polyelectrolytes of zero diameter, for simplicity. The case $D > 0$ leads to similar conclusions.

A contour point of a wormlike polyion is chosen to be the origin O of our Cartesian coordinate system. The contour distance from O is specified by s_1. The vector tangential to the chain at O is constrained to point in the z direction (see Figure 1). The first chain interacts with a second, one contour point of which is fixed at $(R,0,0)$ with the associated tangent vector pointing along the Z_2 axis. The contour distance s_2 is defined with respect to this contour point. The Z_2 axis is skewed at an angle γ with respect to the Z_1 axis. I wish to investigate the average interaction between two nearby sections of the two chains ($-\ell \leq s_1 \leq \ell$, $-\ell \leq s_2 \leq \ell$; ℓ is specified below). They are statistically close to the rod limit, so that they are uniquely specified by the vectors $\varepsilon_1(s_1)$ and $\varepsilon_2(s_2)$ perpendicular to the Z_1 and Z_2 axes respectively. Of course, two long chains may have many domains of nearby interaction like the one shown in Figure 1, but in each case a similar analysis holds.

For the moment it suffices to use the Debye-Hückel approximation. In the continuum limit the average interaction between the two sections can be written as

$$V_{12}/kT = \lambda_B b^{-2} \ll \int_{-1}^{1} ds_1 \int_{-1}^{1} ds_2 \frac{e^{-\kappa r_{12}}}{r_{12}} \gg \tag{1}$$

where $\ll \gg$ denotes the average over the undulations of both sections, b is the linear charge spacing, λ_B the Bjerrum length, T the temperature, k Boltzmann's constant and r_{12} the distance between points s_1 and s_2. To a first approximation, the two sections fluctuate independently, so we have (2)

$$<\epsilon_i^2> = O\left(\frac{s_i^3}{L_p}\right) \qquad\qquad i = 1,2 \tag{2}$$

This can be understood in simple terms: the angular fluctuation of a worm near the rod limit obeys the central limit theorem $<\theta_i^2> \simeq s_i/L_p$ and furthermore $<\varepsilon_i^2> \simeq s_i^2 <\theta_i^2>$ by a geometrical argument. Pythagoras' theorem yields

$$r_{12}^2 = r_{12,0}^2 + O(\epsilon_1^2) + O(\epsilon_2^2) \tag{3}$$

$$r_{12,0}^2 = R^2 + (Z_1(s_1) - Z_2(s_2)\cos\gamma)^2 + Z_2^2(s_2)\sin^2\gamma \tag{4}$$

where $r_{12,0}$ refers to the interaction between charges projected on the respective Z_1 and Z_2 axes, the reference configuration if the polyions were infinitely stiff. We next gauge the impact of the undulatory $O(\epsilon_i^2)$ terms in equation 3.

Relevant configurations for a system without positional order are those for which $V_{12} = O(kT)$ since the Boltzmann factor associated with equation 1 approaches zero rapidly for $V_{12} > > kT$. Owing to the screening factor, we have

$R = O(\kappa^{-1})$ to within logarithmic order. Furthermore, the leading contribution to equation 1 stems from the region $s_1 = O(s_2)$. Since the segments are close to the rod limit, we have to account mainly for separations of the following type in an explicit evaluation of equation 1.

$$r_{12}^2 = O(\kappa^{-2}) + O(s_1^2 \sin^2 \gamma) + O\left(\frac{s_1^3}{L_p}\right) \tag{5}$$

There are then two extreme cases:
   i) $s_1 = O(\kappa^{-1})$; the relevant contribution from the undulation is $O(1/L_p \kappa)$.
   ii) $s_1 = O(\ell)$; the relative undulatory contribution
$= O(\ell L_p^1 \sin^{-2} \gamma)$ and depends on ℓ. In the isotropic phase, we have $\sin \gamma = O(1)$ and the persistence length L_p is the relevant scale i.e. $\ell = O(L_p)$. In the nematic, the deflection length $\lambda = L_p \gamma^2$ is the important characteristic scale (3), i.e. $\ell = O(\lambda)$. Therefore, in both cases the relative magnitude of the last two terms in equation 5 is at most of order unity. I conclude that undulations merely perturb the interaction between two polyions at close separations for solutions without long range positional order. Although a cholesteric solution is positionally ordered, the director undergoes only a gradual twist for the pitch is predicted to be a large mesoscopic scale (4). Hence, cholesteric organization should have virtually no influence on the local interaction displayed in Figure 1.
   Current polyelectrolyte theories of the nematic phase (3,5,6) disregard local undulations which appears to be a reasonable zero-order approximation in view of the above arguments. They agree quite well with experiments on DNA and Xanthan (7-9).

Hexagonal phase

The hexagonal phase has long-range positional order defined by the spacing R of the reference configuration, the hexagonal array of perfectly rigid rods (Figure 2). The spacing is fixed by some external constraint (volume of the sample, imposed osmotic stress, Van der Waals forces, etc.), and is often substantially greater than the Debye radius κ^{-1}. In addition, the undulation amplitude d of the wormlike polyions may also be larger than κ^{-1}, thus modulating the electrostatic screening considerably. On average, the result is an undulation-enhanced interaction. In the previous section, the distance $R = O(\kappa^{-1})$ was determined by a balance of electrostatic forces and thermal motion so an entirely different scenario prevails. A summary of the simple theory developed in reference 10 is given here, with greater emphasis on its limitations.
   In Figure 2 thermal excitations cause the polyions to fluctuate away from the reference configuration which is obviously opposed by their repulsive interaction. An effective way to approach this problem is to posit a spatial distribution function G depending on some variational parameter (the amplitude d in this case), calculate the total Helmholtz free energy A_{tot} self-consistently, and minimize A_{tot} with respect to d keeping the spacing R constrained. We bear in mind that the system is translationally invariant along the long hexagonal axis, the distribution and its derivative are continuous, and G must be sharply peaked at the lattice points otherwise the notion of an hexagonal phase would not be meaningful. In the limit $d < < R$ we may suppose the Gaussian distribution (10)

$$G \sim (\pi d^2)^{-1} e^{-r^2/d^2} \tag{6}$$

to be correct to the leading order, at least for screened or "soft" confinement. In effect, G may be regarded as a Boltzmann factor $\exp{-V_{eff}/\kappa T}$ where the

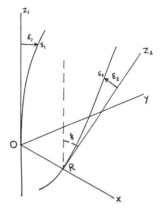

Figure 1.   Two nearby sections of two interacting polyions.

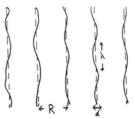

Figure 2.   Two dimensional representation of a   hexagonal lattice of wormlike polyions.

effective self-consistent potential V_{eff} exerted on an infinitesimal segment of a test polyion is expanded to second order in r, the distance from the segment to its respective position in the reference configuration.

Equation 6 asserts that a test polyion is trapped in a fuzzy cylindrical tube. The configurational or undulatory free energy F_{conf} is calculated approximately via a second moment condition implied by equation 6

$$<r^2> = d^2 \tag{7}$$

Helfrich and Harbich (11) presented a fluctuation theory for a worm restricted in such a fashion only

$$A_{conf} = 2^{-2/3} kT \lambda^{-1} \tag{8}$$

In essence, this agrees with a scaling analysis (12) proposed earlier in terms of the deflection length $\lambda \equiv d^{2/3} L_p^{1/3}$ (In order to avoid confusion, it is well to stress that equation 8 applies to soft confinement as expressed by equation 6. In ref. (12) I considered the statistics of a worm trapped in a hard cylindrical pore. In ref. (11) A_{conf} is derived within a second moment restriction and then rescaled geometrically to account for confinement within a hard tube. The predictions of refs. (11) and (12) have been borne out by extensive Monte Carlo simulations (32). A scaling law analogous to equation 8 is verified but of course the numerical coefficient depends on the type of confinement. Note that there is an important difference between soft and hard confinement: in the latter case there is a depletion entropy because the chain cannot cross the cylindrical boundary).

A test polyion can be viewed as an sequence of deflection segments. Each behaves more or less like a cylindrical rod exerting an electrostatic potential $\phi(s)$ at s, the vector distance from the rod axis. In the outer double layer, Debye-Hückel superposition holds so we have a renormalized potential ϕ_R which is the bare ϕ smeared out by the distribution given by equation 6

$$\phi_R(t) = <\phi> = \int drG(r)\phi(t-r) \tag{9}$$

The vector t is defined with respect to the reference axis. Next, the test polyion interacts with six neighbors i.e. there are three pair interactions per test chain. Since the neighboring chains are smeared charge distributions themselves, the electrostatic free energy of interaction per unit length of polyion becomes

$$A_{el}/kT = 3 \nu_{eff} \int dtG(t)\phi_R(R-t) \tag{10}$$

At large distances the polyion may be regarded as a line charge of spacing $1/\nu_{eff}$ (10). An asymptotic evaluation of equations 9 and 10 is fairly straightforward (10)

$$A_{el}/kT = \frac{3(2\pi)^{1/2} \xi_{eff}^2 \exp(-\kappa R + 1/2 \kappa^2 d^2)}{Q(\kappa R)^{1/2}(1 + 1/2 \kappa d^2 R^{-1})} \tag{11}$$

with $\xi_{eff} \equiv \lambda_B \nu_{eff}$. The effect of undulations is marked when the undulatory amplitude d is of the order of the Debye radius.

Minimization of the total free energy $A_{tot} = A_{conf} + A_{el}$ yields a nonlinear expression for d

$$\frac{d^{8/3}e^{\frac{1}{2}\kappa^2 d^2}}{1+\frac{1}{2}\,\kappa d^2 R^{-1}} = \frac{2^{1/3}\lambda_B R^{\frac{1}{2}}e^{\kappa R}}{9(2\pi)^{\frac{1}{2}}\,\xi_{eff}^2\,L_p^{1/3}\,\kappa^{3/2}} \tag{12}$$

and an osmotic pressure (10)

$$\pi_{os} = \frac{2^{1/3}kT}{3^{3/2}\kappa R d^{8/3}L_p^{1/3}} \tag{13}$$

At this stage it is important to list the considerable number of physical approximations implicit in the derivation of equations 12 and 13.
a) $R\kappa >>> d$. An excess of salt is present.
b) $L_p >>> d$. This ensures that the hexagonal lattice exists. There is virtually no folding of the chains. Furthermore, a deflection segment of length $\lambda = L_p^{1/3}d^{2/3} >> d$ has a large aspect ratio so that the orientational fluctuations of order d/λ are small.
c) $\kappa d^3 R^{-2} << 1$. This ensures that the asymptotic analysis of equations 9 and 10 is viable.
d) i) $d << R$; ii) $\frac{1}{2}\kappa^2 d^2 + 3\kappa d + \kappa D \le \kappa R$ (D = bare diameter of the polyion). The first inequality may seem adequate enough to guarantee the validity of the Gaussian ansatz. Yet, the tail of the Gaussian is weighted heavily in the average interaction. If we take due care of a cut-off at R, we derive the second more restrictive condition. Then, a self-consistent theory neglecting certain correlations is reasonable. In a similar vein, the neglect of hexagonal symmetry in the isotropic distribution G is justified.
e) $\kappa\lambda >> 1$. The deflection segment has the electric properties of a charged rigid rod.
f) $(D^{\frac{1}{2}}+\kappa^{-\frac{1}{2}})/\kappa^{\frac{1}{2}}\lambda > d/\lambda$. Electrostatic twist is negligible (*3*). The polyions are basically parallel. Orientation-translation coupling can be disregarded. Still, though very weak, undulations play an important role.
g) $R \ge D + 2d + 2\kappa^{-1}$. Counterion fluctuations do not couple to undulations.
h) $\lambda F_{el}/kT \le 1$ for $\kappa d \ge 1$. The electrostatic interaction of an independently fluctuating deflection segment is small enough to circumvent a virial expansion (*13*).

Equations 12 and 13 have the following features.
1) For $\kappa d << 1$, the osmotic pressure tends to a previous result (*14*) for an hexagonal array of rigid cylindrical polyions i.e. our reference configuration

$$\pi_{os,0} = \left(\frac{(6\pi)^{\frac{1}{2}}\,\xi_{eff}^2\kappa^{\frac{1}{2}}}{\lambda_B}\right)\frac{e^{-\kappa R}}{R^{3/2}} \tag{14}$$

2) For $\kappa d << 1$, the first-order correction to equation 14 reads

$$\pi_{os} = \pi_{os,0} + \pi_{os,1} \tag{15}$$

where $\pi_{os,1} << \pi_{os,0}$ and $\pi_{os,1} \sim \exp(-\kappa R/4)$. This gradual decay agrees with a qualitative sketch given by Selinger and Bruinsma (*13*), provided we neglect their nematic interaction in accordance with condition f. An enhanced decay length was first proposed by Podgornik and Parsegian (*15*) but for a Gaussian random coil enclosed in a tube. This calculation is valid when the step length is smaller than the tube diameter which is the reverse of the case focused on here

$(L_p > > R)$. Note that an enhanced decay length shows up within a perturbation that is weak.

3) The case of experimental interest is $\kappa d = O(1)$ for which no simple power or decay law can be derived. Recall that in view of restrictions c and d, we cannot take the limit of dominating exponentials in equation 12 which would lead to $d^2 \approx 2R/\kappa$.

4) Similarly, we cannot address the approach to the uncharged hexagonal phase which is strongly fluctuating. In that case, $d = O(R-D)$ and $\pi_{os} \approx kT L_p^{-1/3}(R-D)^{-8/3}$ as discussed at length by Selinger and Bruinsma (13).

5) The list of restrictions (a-h) looks awesome. Still, as a rule, the hexagonal gels formed in osmotic stress experiments (DNA (16), tobacco mosaic virus (TMV, 14, 17), muscle filament (14,18) satisfy these requirements for the polyions are relatively thick, stiff and highly charged. For a detailed comparison of theory with experiment, the reader should consult ref. (10). A Lindemann melting rule has also been formulated with the help of equation (12) (T. Odijk, Europhysics Lett., in press). It appears to explain the stability of the hexagonal phase of DNA (16).

Van der Waals forces

Stability theories of the DLVO type have been developed for hexagonal lattices of rigid polyelectrolytes by various authors (19-21). The work of Parsegian and Brenner (21) inspired Millmann et al (17) to reinvestigate the classic experiments of Bernal and Fankuchen (22) on gels of TMV. How do undulations influence the interpretation of these measurements? To begin with, we must keep in mind that requirements a-h of the previous section are rather restrictive when attractive forces are taken into consideration. In particular, I disregard entirely configurations like those shown in Figure 3 in which chains adhere. If adhesion does occur, the Gaussian distribution evidently becomes a poor approximation. Given these limitations, the first problem to be faced is the possible effect of undulations on the Van der Waals interaction itself.

The bare Van der Waals interaction per unit length between two parallel rigid cylinders is known in terms of a hypergeometric function (23) but physical insight is afforded by studying the expansions at small and large separations (24,25). For an hexagonal lattice we have

$$A_W/kT = -\frac{HD^{1/2}}{8.2^{1/2}(R-D)^{3/2}} \left[1 - \frac{2(R-D)}{D} + \ldots \right] \qquad R-D \ll D \qquad (16)$$

$$A_W/kT = -\frac{9\pi H}{128D} \left(\frac{D}{R} \right)^5 \left[1 + \frac{25}{16}\frac{D^2}{R^2} + \frac{31.9}{16}\frac{D^4}{R^4} + \frac{150.7}{64}\frac{D^6}{R^6} + \ldots \right] \qquad (17)$$

$$R \geq 3/2D$$

where H is the Hamaker constant scaled by kT. I now approximate the undulatory average of the lead term of equation 16 by noting that those fluctuations can be neglected in a zero-order analysis, that are transverse to the vector connecting the two axes of the deflection segments in the reference configurations. It is convenient to introduce the relative change J in A_W

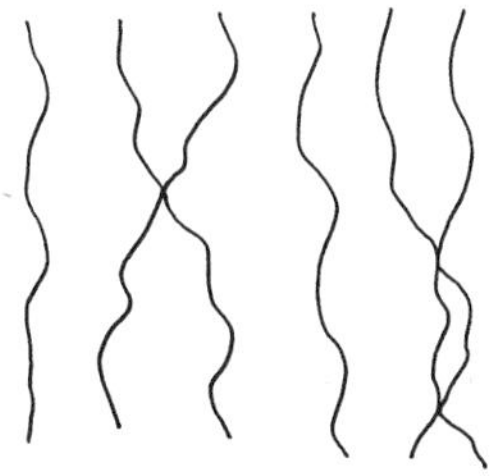

Figure 3.   Forbidden adhesive states.

$$J \equiv E^{3/2} \ll (E+X_1+X_2)^{-3/2} \gg$$

$$\simeq \pi^{-1}d^{-2}E^{3/2} \int_{-C}^{C} \int dX_1 dX_2 e^{-(X_1^2+X_2^2)/d^2}(E+X_1+X_2)^{-3/2} \qquad (18)$$

with $E \equiv R\text{-}D$, and X_1 and X_2 the longitudinal positions of the two segments. Here and below, C denotes a formal cut-off which temporarily avoids a divergence at contact. It is assumed that the chains never enter their primary minimum so the real distribution is zero at zero separation. On introducing polar coordinates $X_1 = r\cos\phi$, $X_2 = r\sin\phi$, $W \equiv r^2$, and the Legendre function $P_{1/2}(z)$, we obtain (31)

$$J = d^{-2} \int_0^C dW \ e^{-W/d^2} P_{1/2} \left[\left[\frac{E^2}{E^2-2W} \right]^{1/2} \right] (1-\frac{2W}{E^2})^{-3/4} \qquad (19)$$

For $z \geq 1$, the Legendre function is approximated quite well by $z^{1/2}$.

$$J \approx d^{-2} \int_0^C dW \ e^{-W/d^2} \left[1-\frac{2W}{E^2} \right]^{-1} \qquad (20)$$

$$\approx -\frac{1}{4}E^2 d^{-2}\log\left[1-\frac{4d^2}{E^2} \right] \qquad (2d<E) \qquad (21)$$

The second form is chosen because it has the correct Taylor expansion for small d/E and a logarithmic divergence is expected at close packing when $d = O(E)$ and the Gaussian approximation ultimately breaks down. For $d < 6E = 6(R\text{-}D)$, J deviates from unity by six percent at most. Similar considerations apply to equation 17, an undulatory correction being negligible since $d^2 <<< R^2$.

In the application to TMV gels, the analysis proceeds by investigating the total free energy $A_{tot}(d,R) = A_{conf} + A_{e\perp} + JA_w$ numerically as a function of R. The variational amplitude d is determined by minimizing A_{tot} for each spacing R. There are two solutions, the one for higher d being unstable. For $R \geq (3/2) D$, the factor J is set equal to unity. Since the function $A_{tot}(R)$ is very sensitive to the values of H and L_p which are not known precisely, the scaled Hamaker constant H is adjusted so as to let the spacing R corresponding to the *secondary* minimum, conform to the experimental spacing R_{exp}. The latter were measured for hexagonal TMV gels as a function of the electrolyte concentration by Millman et al (17) after long equilibration under zero external stress. Representative values of the microscopic parameters are (17):TMV diameter D=18 nm, length L=300 nm, real charge density $b^{-1}=14$ nm^{-1}, Bjerrum length $\lambda_B=0.71$ nm. The scaled Hamaker constant H is estimated to range from about unity to two for proteins at room temperature (21,25). The quantity ξ_{eff} is calculated using the Philip-Wooding solution (26) to the Poisson-Boltzmann equation. In electron micrographs, fluctuations of TMV from the straight rod configuration typically amount to 5 nm so the persistence length L_p should be of order 10^4 nm from equation 2.

A comparison with experiments (*17*) is given for three ionic strengths.

i) C=0.87 M. The spacing R_{exp} is about 20 nm or a bit higher (17). The existence of a feasible secondary minimum for the free energy $F_p \equiv LF_{tot}/kT$ of a TMV polyion is established for a plausible range of values of L_p. Several combinations of parameters at the minimum are compiled in Table I. Only a very narrow range of H is allowed for each value of L_p, both here and below.

Table I. Combinations of Parameters at the Minimum F_p at C = 0.87 M

Lp(nm)	3.10^4		10^4		3.10^3
H	0.6	0.5	0.7	0.6	0.8
R(nm)	19.8	20	19.6	19.9	19.6
Fp	-1.3	-1.44	-4.91	-0.59	-2.4
κd	1.42	1.26	1.20	1.45	1.34

From a theoretical point of view undulations cannot be dismissed for stable gels of TMV ($\kappa d = O(1)$). The influence of undulations on the Van der Waals interaction is significant (J=1.1-1.2) because confinement entropy, and electrostatic and dispersion forces, are delicately balanced at the secondary minimum. The H values are compatible with independent estimates (*21,25*).

ii) C=0.096 M. The spacing R_{exp} is about 28 nm with a fairly large variance. Stability analysis results in Table II.

Table II. Combinations of Parameters at Spacing R_{exp} of 28 nm and C = 0.096 M

Lp(nm)	3.10^4	10^4	3.10^3
H	6	7	10
R(nm)	26	25.5	25
Fp	-1.43	-0.70	-2.23
κd	1.29	1.24	1.22

The values of H are an order of magnitude greater than those in case (i) and too large to be realistic. If the present Gaussian approximation is taken seriously, an additional attractive force needs to be postulated.

iii) C=0.01 M; $R_{exp} \approx 35$ nm. The scaled Hamaker constant H needs to be of order 10^2, independent of L_p. The existence of another attractive interaction is unambiguous. I remark that the existence of adhesive states (Figure 3) is extremely unlikely in view of the relatively small undulations (d/R=0.05 although $\kappa d = O(1)$).

I conclude that undulation theory explains the stability of TMV gels solely at high ionic strength. At low ionic strength, we can state that adhesive states should not occur because the undulations would have to surmount a huge electrostatic barrier. It is now thought that hydration forces may be attractive under the right conditions (*27,28*). In fact, if TMV were to have a long range hydrophobic attraction, decaying exponentially and similar to the one displayed in Figure 3 of ref. (*28*), it would overwhelm the usual Van der Waals force at low ionic strength, but not at 1 M. For the moment, this is mere speculation. Doubtless, further investigation is needed on the following topics: a) a rigorous self-consistent theory incorporating dispersion forces, possibly of the Lifshitz type; b) in several of the calculations for TMV, A_{tot} becomes *positive* beyond the secondary minimum because undulation enhancement is powerful; this raises the possibility of a tertiary minimum in the strongly fluctuating limit $(d = O(R^{1/2}\kappa^{-1/2}))$ to $O(R)$; for the theory of uncharged adhering polymers, see refs. (*11*), (*29*) and (*30*)); c) the quantitative analysis of newly postulated forces; d) dynamic stability.

Literature Cited

1. Durand, D.; Doucet, J.; Livolant, F. *J. Phys. II France* 1992, *2*, 1769.
2. Odijk, T.; Mandel, M. *Physica* 1978, *93A*, 298.
3. Odijk, T. *Macromolecules* 1986, *19*, 2313.
4. Odijk, T. *J. Phys. Chem.* 1987, *91*, 6060.
5. Vroege, G.J. *J. Chem. Phys.* 1989, *90*, 4560.
6. Khokhlov, A.R. in *"Liquid Crystallinity in Polymers"*, ed. A. Ciferri, VCH, New York, 1991.
7. Strzelecka, T.; Rill, R.L. *Macromolecules* 1991, *24*, 5124.
8. Inatomi, S.; Jinbo. Y.; Sato, T.; Teramoto, A. *Macromolecules* 1992, *25*, 5013.
9. Vroege, G.J.; Lekkerkerker, H.N.W. *Rep. Prog. Phys.* 1992, *55*, 1241.
10. Odijk, T. *Biophys. Chem.* 1993, *46*, 69.
11. Helfrich, W.; Harbich, W. *Chemica Scripta* 1985, *25*, 32.
12. Odijk, T. *Macromolecules* 1983, *16*, 1340.
13. Selinger, J.V.; Bruinsma, R.F. *Phys. Rev.* A 1991, *43*, 2910, 2922.
14. Millman, B.M.; Nickel, B.G. *Biophys. J.* 1980, *32*, 49.
15. Podgornik, R.; Parsegian, V.A. *Macromolecules* 1990, *23*, 2265.
16. Podgornik, R.; Rau, D.C.; Parsegian, V.A. *Macromolecules* 1989, *22*, 1780.
17. Millman, B.M.; Irving, T.C.; Nickel, B.G.; Loosley-Millman, M.E. *Biophys. J.* 1984, *45*, 551.
18. Millman, B.M.; Wakabayashi, K.; Racey, T.J. *Biophys. J.* 1983, *41*, 259.
19. Elliott, G.F. *J. Theor. Biol.* 1968, *21*, 71.
20. Brenner, S.L.; McQuarrie, D.A. *Biophys. J.* 1973, *13*, 301.
21. Parsegian, V.A.; Brenner, S.L. *Nature* 1976, *259*, 632.
22. Bernal, J.D.; Fankuchen, I. *J. Gen. Physiol.* 1941, *25*, 111.
23. Bouwkamp, C.J. *Kon. Ned. Akad. Wetenschap* 1947, *50*, 1071.
24. Sparnaay, M.J. *Rec. Trav. Chim. Pays-Bays* 1959, *78*, 680.
25. Mahanty, J.; Ninham, B.W. *"Dispersion Forces"*, Academic, New York, 1976.
26. Philip, J.R.; Wooding, R.A. *J. Chem. Phys.* 1970, *52*, 953.
27. Rau, D.C.; Parsegian, V.A. *Biophys. J.* 1992, *61*, 24.6
28. Christenson, H.K.; Claesson, P.M.; Parker, J.L. *J. Phys. Chem.* 1992, *96*, 6725.
29. Maggs, A.C.; Huse, D.A.; Leibler, S. *Europhys. Lett.* 1989, *8*, 615.

30. Lipowsky, R. *Phys. Scripta* 1989, *T29*, 259.
31. Gradshteyn, I.S.; Ryzhik, I.M. *"Table of Integrals, Series and Products"*, Academic, London, 1980.
32. Dijkstra, M.; Frenkel, D; Lekkerkerker, H.N.W. *Physica A* 1993, *193*, 374.

RECEIVED August 6, 1993

Chapter 8

Quasi-elastic Light Scattering by Polyelectrolytes in Buffers of Low Ionic Strength

Narinder Singh[1], Kenneth S. Schmitz, and Yueying Y. Gao[2]

Department of Chemistry, University of Missouri—Kansas City, Kansas City, MO 64110

The q-dependent apparent diffusion coefficient, $D_{app}(q)$, for translational diffusion of hard spheres in the *weak interaction limit* is discussed in detail. In this theory the small ion-polyion coupled mode equation is solved exactly for added electrolyte and salt-free solutions of a monodispersed solution of charged spheres. Electrolyte dissipation and hydrodynamic interactions are included in an *ad hoc* manner. The model is applied to the systems of bovine serum albumin (BSA) obtained in our laboratory and polystyrene latex spheres (PLS) reported in the literature. It is concluded that the q = 0 theories are valid for the BSA data but not for the PLS data.

Interparticle interactions greatly influence the dynamics of macromolecules in solution and colloidal suspensions. Perturbation methods can provide information regarding the nature of these interactions, but only in their relaxation from a sometimes highly disruptive state to the so-called equilibrium state of the system. It is well-known that the viscosity of charged macroions is shear-rate dependent. Giordano *et al.* (*1*), for example, reported that the viscosity of bovine serum albumin (BSA) determined with a low shear rotating cylinder viscometer was about 10 times larger that those obtained from the capillary flow method. The more subtle weak interparticle interactions may, therefore, go entirely undetected unless one can somehow examine these systems with a sensitive, non perturbation method. Quasi-elastic light scattering (QELS) is such a non perturbative method for studying the solution dynamics of macromolecules. In the QELS technique, temporal fluctuations in the scattered light intensity are monitored, from which molecular information is deduced from the autocorrelation function of the scattered light intensity.

The focus of this communication is on the "ordinary" polyion behavior that results from the weak electrostatic coupling between the various charged species present in the solution. Attention is given to the q-dependence of the apparent diffusion coefficient, $D_{app}(q)$, for the small ion-polyion coupled mode (CM) system.

[1]Current address: Department of Biochemistry, University of Kansas Medical Center at the VA Medical Center, 4801 Linwood Boulevard, Kansas City, MO 64128
[2]Current address: Department of Chemistry, Peking University, Beijing 100871, People's Republic of China

0097–6156/94/0548–0098$06.00/0

Generalized Diffusion Equation for Interacting Hard Sphere Systems

Consider an electrically neutral volume element in the solution located at $\mathbf{r}$ with a uniform particle concentration for the jth component represented as $<n_j>_u$ (particles/mL). The combined Fick's first and second laws of diffusion for the temporal evolution of a *concentration fluctuation* for the jth component from the *uniform distribution* $<n_j>_u$, i. e., $\Delta n_j(\mathbf{r}, t) = n_j(\mathbf{r}, t) - <n_j>_u$, is given by,

$$\left\langle \frac{\partial \Delta n_j(\mathbf{r}, t)}{\partial t} \right\rangle = -\nabla \bullet \left\langle j_j \right\rangle = \nabla \bullet \left\langle \underline{\mathbf{D}_j} \bullet \left[\nabla \Delta n_j(\mathbf{r}, t) + \beta \nabla \phi(\mathbf{r}, t) \right] \right\rangle \tag{1}$$

where $<j_j>$ is the average particle flux of the jth component, $\underline{\mathbf{D}_j}$ is the diffusion tensor that couples the jth component with the other particles in the complex fluid through *kinetics related* interparticle interactions (viz., dissipation forces), $\beta = 1/k_B T$ where k_B is Boltzmann's constant and T is the absolute temperature, and $\phi(\mathbf{r}, t)$ is the *direct interaction potential* . Before proceeding to the generalized expression for small ion-polyion coupled modes, it is necessary to review the approximations usually employed to simplify equation 1 that provides tractable expressions for interpreting experimental data.

The solution to equation 1 is generally converted to an expression for the average quantities, with the approximation,

$$\nabla \bullet \left\langle \underline{\mathbf{D}_j} \bullet \nabla \right\rangle = \left\langle D_j \right\rangle \nabla^2 \tag{2}$$

Substitution of equation 2 into equation 1 may limit the range of concentrations to which the expression can be applied. For example, the hydrodynamic interaction between two spheres of radius R_S separated by a distance r_{ij} is expressed in terms of power laws in the relative parameter R_S/r_{ij}, viz. $(R_S/r_{ij})^n$. Carter and Phillies (2) have shown that using the right hand side of equation 2 is restricted to $n < 7$ since the divergence of higher powers is not zero.

The average of the second term in equation 1 is expressed as,

$$\beta \nabla \bullet \left\langle \underline{\mathbf{D}_j} \bullet \nabla \phi(\mathbf{r}, t) \right\rangle = \nabla \bullet \left\langle \frac{\nabla \phi(\mathbf{r}, t)}{f_m} \right\rangle \cong \left\langle \frac{1}{f_m} \right\rangle \nabla \bullet \left\langle \nabla \phi(\mathbf{r}, t) \right\rangle \tag{3}$$

where the inverse friction tensor for the jth particle is represented by the mutual friction factor f_m. Anderson and Reed (3) examined the consequences of the right-most approximation in equation 3 for a one-dimensional system. Using only the hard sphere and long range hydrodynamic interaction potentials, they reported that, as the two spheres approached each other, the right-most average defined in equation 3 led to an increased viscous force such as to prevent the spheres from touching. Hence in their analysis there is no direct hard sphere repulsion interaction term in the evaluation of the second virial coefficient.

Generalized Diffusion Equation for Small Ion-Polyion Coupled Modes: Weak Electrostatic Interaction Limit

Lin, Lee, and Schurr (4) considered jth component of charge $Z_j e$ (Z_j includes both magnitude and sign, and e is the magnitude of the electron charge) diffusing under the influence of an electrical potential $\phi_{elec}(\mathbf{r}, t)$ in accordance with the equation,

$$\frac{\partial \Delta n_j(\mathbf{r}, t)}{\partial t} = D_j^0 \nabla^2 \Delta n_j(\mathbf{r}, t) + Z_j e \beta \langle n_j \rangle_u D_j^0 \nabla^2 \phi_{elec}(\mathbf{r}, t) \tag{4}$$

where D_j^0 is in the their theory the infinite dilution limit of the diffusion coefficient of the jth species in the absence of electrostatic coupling with the other particles. They assumed that the electrical potential obeyed the Poisson equation, which generalized to a system of N types of ionic species is,

$$\beta \nabla^2 \phi_{elec}(\mathbf{r}, t) = -4\pi \lambda_B \sum_{i=1}^{N} Z_i \Delta n_i(\mathbf{r}, t) \tag{5}$$

where $\lambda_B = e^2 / \epsilon k_B T$ is the Bjerrum length and ϵ is the dielectric constant of the medium. Note that $\phi_{elec}(\mathbf{r}, t)$ .arises from particle fluctuations away from the <u>uniform</u> concentration distribution. Quasi-elastic light scattering measurements do not monitor concentrations in real space, but in momentum space, hence one substitutes $\Delta n_j(\mathbf{r}, t) \to \Delta n_j(\mathbf{q}, t)$ and $\nabla^2 \Delta n_j(\mathbf{q}, t) \to -q^2 \Delta n_j(\mathbf{q}, t)$ into equation 4, where the scattering vector q is defined as $q = (4\pi \tilde{n} / \lambda_o) sin(\theta/2)$ for the scattering angle θ, wavelength of incident light λ_o, and index of refraction $\tilde{n}$. The resulting equation in q-space in matrix form is,

$$\frac{\partial \Delta \mathbf{n}(\mathbf{q}, t)}{\partial t} = -\underline{\Omega}(q) \bullet \Delta \mathbf{n}(\mathbf{q}, t) \tag{6}$$

where $\underline{\Omega}(q)$ is the frequency matrix whose elements are given by,

$$[\underline{\Omega}(q)]_{ij} = D_j \left[q^2 \delta_{ij} + \frac{Z_i}{Z_j} \kappa_j^2 \right] \tag{7}$$

where δ_{ij} is the Dirac delta function and the *partial* screening parameter κ_j is defined by $\kappa_j^2 = 4\pi \lambda_B Z_j^2 < n_j >_u$. Note that D_j occurs in equation 7 rather than D_j^0., in anticipation of later inclusion of hydrodynamic and electrolyte dissipation effects.

$D_{app}(q)$ for Polyions in the Presence of Added Electrolyte. It is assumed that there are three species present in the solution: the polyions (component 1); the counterions (component 2) and the coions (component 3). The frequency matrix for this system is therefore a 3x3 matrix whose secular cubic equation has been solved exactly using the Cardan solution. The coefficients of the nth power term, c_n, in the secular equation are: $c_1 = -(D_1 + D_2 + D_3)q^2 - (D_1 \kappa_1^2 + D_2 \kappa_2^2 + D_3 \kappa_3^2)$; $c_2 = (D_1 D_2 + D_2 D_3 + D_3 D_1)q^4 + [D_1 D_2(\kappa_1^2 + \kappa_2^2) + D_2 D_3(\kappa_2^2 + \kappa_3^2) + D_3 D_1(\kappa_3^2 + \kappa_1^2)]q^2$; and $C_3 = -D_1 D_2 D_3 q^6 - D_1 D_2 D_3 \kappa_{tot}^2 q^4$, where $\kappa_{tot}^2 = \kappa_1^2 + \kappa_2^2 + \kappa_3^2$ is the total screening parameter. The apparent diffusion coefficient for the k+1 root is (5). ,

$$D_{k+1}(q) = \frac{1}{q^2}\left[2\left(\frac{-a^3}{27}\right)^{1/6}cos\left(\frac{\psi + 2\pi k}{3}\right) - \frac{c_1}{3}\right] \qquad (k = 0,1, \text{ or } 2) \qquad (8)$$

where $a = c_2 - (c_1^2/3)$, $b = (2c_1^3/27) - (c_1c_2/3) + c_3$, and $cos(\psi) = -b/[2(-a/3)^{1/2}]$. The polyion diffusion mode is the root for which $k = 1$. Representative calculations for this model are given in Table I.

$D_{app}(q)$ for Mixed Polyion System with No Added Salt. The development in the previous section can be directly applied to a system of mixed polyions with no added salt, where components 1 and 2 represent the polyions and component 3 represents the counterions of concentration, $C_3 = - (Z_1C_1 + Z_2C_2)/Z_3$ (24). Representative results are given in Table II. The roots to the cubic equation that are associated with the polyion diffusion modes ($k = 1$ and 2, or roots R_2 and R_3, respectively) were determined from the apparent diffusion coefficient obtained under high salt-low charge conditions (top row in Table II).

Single Polyion Species with No Added Salt. The frequency matrix for a single macroionic species with no added electrolyte is a 2x2 matrix. From the charge neutrality condition one has the relationship $\kappa_2^2 = -Z_2\kappa_1^2/Z_1$. $D_{app}(q)$ for the polyion is obtained from the quadratic expression, where the "+" root corresponds to the polyion as determined from the high salt-low charge conditions.

Series Expansion Representation of $D_{app}(q)$

Petsev and Denkov (7) represented their $D_{app}(q)$ versus q plots for PLS as a power series expansion to the 4th power in q,

$$D_{app}(q) = D_m + Aq^2 + Bq^4 \qquad (9)$$

As might be the case for premature truncation of the series expansion, the numerical values of D_m, A, and B can adequately fit the data profile but may not have physical significance in regard to the values of molecular parameters calculated from an appropriate theory. Equation 8 has previously been expanded to the q^6 term (5, 8). Hence model dependent expressions for D_m (see next section), A, and B are available to assess the "curve fitted" values for these coefficients. Such a comparison is delayed to the discussion section. Unfortunately the mathematical expression are too long to be reproduced in this short communication, so the interested reader is directed to the literature for the mathematical forms of A and B.

Linearized Form of $D_{app}(q = 0)$

For the special case $Z_2 = -Z_3$ and $D_s = D_2 = D_3$, where D_s is the small ion diffusion coefficient, the q = 0 limit for the expression in equation 8 is the Lin *et al.* (4) result,

$$D_{app}(q = 0) = \frac{1}{2}\left[D_1 + D_s + \frac{1 - x(1+y)}{1 + x(1+y)}(D_1 - D_s)\right] \qquad (10)$$

where $x = D_s/D_1Z_1$, and $y = 2C_s/Z_1C_1$. We distinguish between added electrolyte, C_{add}, and the counterions in our definition of $2C_s$, viz., $2C_s = 2C_{add} + |Z_1|C_1$.

Table I. $D_{app}(q)^a$ for Selected Values of q for a Spherical Macroion in the Presence of Added Electrolyte

$C_1 \times 10^6$	Z_1	$D_1 \times 10^8$	$C_3 \times 10^4$	q (10^5 cm^{-1})			
(M)		(cm²/s)	(M)	1.0	1.5	2.0	2.5
50.0	500	2.0	0.0001	668.0	668.0	667.0	667.0
50.0	500	2.0	1.0	664.0	664.0	664.0	664.0
0.01	50	2.0	1.0	2.2	2.2	2.2	2.2
0.01	50	60.0	0.0001	138.0	95.4	80.1	72.9
0.01	50	60.0	1.0	66.6	66.0	65.3	64.6
0.01	150	20.0	0.0001	242.0	124.0	79.3	58.3
0.01	150	20.0	1.0	83.0	77.4	71.0	64.6
0.10	150	20.0	1.0	195.0	182.0	166.0	149.0

a - units of 10^{-8} cm²/s

Table II. $D_{app}(q)^a$ for Selected Values of q for Mixed Macroions in Salt-free Solutions

$D_1{}^a$	Z_1	$C_1{}^b$	$D_2{}^a$	Z_2	$C_2{}^b$	R_2/q^2	R_3/q^2
1.0	1	0.01	10.0	1	0.01	1.0	10.0
2.0	500	0.01	20.0	150	0.01	11.3	369.0
2.0	500	0.01	20.0	150	1.0	2.2	1170.0
2.0	500	1.0	20.0	150	0.01	19.8	660.0
2.0	150	0.01	20.0	500	0.01	2.16	1340.0
2.0	150	0.01	20.0	500	1.0	2.00	1670.0
2.0	150	1.0	20.0	500	0.001	10.4	447.0

a - units of 10^{-8} cm²/s at a scattering vector $q = 1 \times 10^5$ cm⁻¹;   b- units of 10^{-6} M

Many experiments are carried out under relatively dilute solution conditions such that $D_{app}(q = 0)$ appears to be a linear function of the polyion volume fraction ϕ_1. Equation 10 in the limit $y \gg 1$ yields the desired linearized expression,

$$D_{app}(q = 0) \cong D_1^0\left(1 + A_d\frac{Z_1^2 C_1}{2C_s}\right) = D_1^0\left[1 + \left(A_d\frac{Z_1^2}{2C_s}\frac{3}{4\pi R_S^3}\frac{1000}{N_A}\right)\phi_1\right] \quad (11)$$

where N_A is Avogadro's constant, $A_d = (D_S - D_1)/D_S$ is the dynamic attenuation term due to the finite response time of the small ions to the movement of the polyion.

The Cell Model and the Composite Diffusion Coefficient.

Macroionic solutions and suspensions are multicomponent systems that involve complex interparticle interactions between the solute and solvent particles. No theory exists, therefore, that takes all of these effects into consideration under arbitrary experimental conditions. We therefore resort to an *ad hoc* construction of a composite diffusion coefficient based on the assumption that in the linearized approximation these additional effects are additive.

Hydrodynamic Interaction. Hydrodynamic interactions differ from other types of interparticle interactions in the sense that they can be *reflected* off of the other particles in the solution. A hydrodynamic wake originating from the "test" particle can therefore reflected off of other particles and thereby affect its own motion. Hydrodynamic interactions, therefore, are *indirect* interparticle interactions. These interactions are expressed as a multiplicative factor in the diffusion coefficient, i.e., $D_1(q) = D_1^0 H_{11}(q)$, where D_1^0 is the infinite dilution value of D_1 (*9, 10*).

Electrolyte Dissipation. Electrolyte dissipation refers to the retardation effect on the polyion motion due to the instantaneous distortion of the surrounding ion atmosphere as the macroion moves through the medium. If hydrodynamic and electrolyte dissipation contributions are not coupled, then the infinite dilution expression for the polyion diffusion coefficient is $D_1^0 = k_B T / (f_{SE} + f_{elec})$ where $f_{SE} = 6\pi\eta R_S$ is the usual Stokes-Einstein friction factor for hard spheres and f_{elec} accounts for electrostatic dissipation effects. Expressions have been obtained for f_{elec} for a charged sphere without (*11-14*) and with (*15*) hydrodynamic interaction between the small ions and the polyion.

Direct Macroion-Macroion Interactions in Strong Interaction Limit. It is well-known that D_1 for a system of strongly interacting macroions exhibits a q-dependence due to the highly ordered nature of the macroions in solution or colloidal suspension. The q-dependence is manifested in the macroion-macroion partial structure factor $S_{11}(q)$. The inclusion of both the indirect hydrodynamic and direct interactions between the macroions yields the following form for $D_1(q)$ (*9, 10*)

$$D_1(q) = \frac{D_1^0 H_{11}(q)}{S_{11}(q)} \quad (12)$$

In the $q = 0$ limit the osmotic susceptibility and the solution structure factor are related through the pairwise direct interaction potential, where it is easily shown through the second virial coefficient that,

$$\frac{1}{RT}\left(\frac{\partial \pi}{\partial C_1}\right)_{\mu,T} = \frac{1}{S_{11}(q=0)} \tag{13}$$

If the interparticle interaction between the macroions is composed of a hard sphere and weak screened Coulombic interaction, then to the first power in the concentration,

$$\frac{1}{S_{11}(q=0)} \cong 1 + \left(8 + \frac{Z_1^2}{2C_s}\frac{3}{4\pi R_S^3}\frac{1000}{N_A}\right)\phi_1 \tag{14}$$

The first term in the parentheses of equation 14 results from the hard sphere potential and the second term from the screened Coulombic interaction in which the exponential form of the long range pair distribution function is expanded to only the first two terms.

Dilemma of the Electrostatic Interactions. In comparison of equations 11, 13, and 14 one might conclude that the two electrolyte dependent terms are the same if $A_d = 1$. These two terms, however, result from two mutually exclusive model assumptions regarding the role of the electrolyte ions. In equation 11 the electrolyte ions play an *active role* in the dynamics of the macroions through the coupled terms in the Poisson equation, whereas in equation 14 the electrolyte ions act in a *passive manner* by simply screening the interaction between the dynamically coupled macroions. It is our opinion that these two terms, although numerically identical if $A_d = 1$, have two distinct origins in the scheme of interparticle interactions of macroionic solutions and suspensions. The manner in which the separate origins of active and passive contributions of the small ions affects the expression of the mutual diffusion coefficient is discussed in the following section.

The Cell Model and the Modified Form of $D_{app}(q = 0)$

Partial support for the supposition of separate contributions of polyion-polyion and polyion-small ion to $D_{app}(q = 0)$ is found in the cell model of Imai and Mandel (*18*) and of Penfold and co-workers (*19*). In this approach the cell is assumed to contain one polyion and a sufficient number of electrolyte ions to be electrically neutral. Penfold and co-workers (*19*) explicitly considered three types of electrostatic interactions: 1) the macroion interactions with the counterions inside the cell; 2) the interaction between macroions in neighboring cells, and 3) the interactions between electrolyte ions of neighboring cells. They found that the latter interactions are negligible in comparison with the first two types of interactions. Hence D_1 in equation 7 is assumed to be the same form as in equation 12, viz.,

$D_1 \to D_1(q) = D_1^0 H_{11}(q) / S_{11}(q)$. Within the context of the Penfold *et al.* study, we interpret this substitution in terms of polyion-polyion interactions in different cells and equation 7 as the frequency matrix for macroion-electrolyte coupled dynamics within the cell. If these contributions are additive as supposed, then the mutual diffusion coefficient, $D_m = D_{app}(q = 0)$, for the cell model is,

$$D_m = D_1^0\left[\frac{1}{RT}\left(\frac{\partial \pi}{\partial C_1}\right)_{\mu,T} + A_d\frac{Z_1^2 C_1}{2C_s}\right]H_{11} \cong D_1^0(1 + \lambda\phi_1) \tag{15}$$

where D_l^o includes electrolyte dissipation effects. This form for D_m is precisely the same as that of Imai and Mandel (*18*) for the cell model if one sets $A_d = 1$. In view of the above development the generalized form of the coefficient λ is,

$$\lambda = 8\delta_{hs} + \left(\delta_{11} + A_d\right)\frac{Z_l^2}{2C_s}\frac{3000}{4\pi R_s^3 N_A} - k_h \tag{16}$$

where $\delta_{hs} = 1$ if hard sphere interactions are important or $\delta_{hs} = 0$ if there are no hard sphere contributions (such as in the Anderson-Reed model), $\delta_{11} = 1$ if the concentration coefficients in equations 11 and 14 have different origins (cell model) or $\delta_{11} = 0$ if these equations have a common origin, and k_h accounts for the hydrodynamic interactions *between the macroions*. The numerical value of k_h is model dependent. For example, Carter and Phillies (*17*) obtained the value $k_h = 8.898$ for uncharged hard spheres where divergent terms (cf. equation 2) and the effect of solvent back flow are included.

Experimental

Many QELS studies in the literature rely on data obtained in other laboratories on similar samples for the interpretation of QELS data. In order to better assess the relative importance of the various contributions to the diffusion coefficient it is important to perform several complementary experiments on the same preparation of the sample. We summarize, therefore, the studies of Singh (*20*) on well-defined BSA preparations and the complementary measurements of QELS, total intensity (I_{tils}), osmotic susceptibility, and viscosity.

Preparation and characterization of BSA monomers. The BSA was purchased from Sigma Chemical Company in St. Louis (Lot No. A-6003, Fraction V, essentially fatty acid free). Gel permeation chromatography (12 in long column, Bio-Gel P-200, Tris-HCl buffer, pH 7.2, 5 °C) separated the monomers from the dimers and trimers as monitored by absorbance of the elution and later verified by mini-slab gel electrophoresis. The running gel was 30% (w/v) acrylamide with 0.8% (w/v) bis-acrylamide in Tris-HCl pH 8.8 buffer with 1% (w/v) SDS. The electrophoresis was performed initially with 60 volts until the dye reached the running gel, after which 100 volts was applied for 8-12 hours. The gels were stained with coomassie blue for 8-10 hours and then destained with a mixture of 2-propanol and acetic acid. The gels were then photographed with a Polaroid MP-4 Land camera for future reference.

Preparation of the dialysis bags. The BSA samples used in the osmometry and light scattering experiments were exhaustively dialyzed against the appropriate solvent. Pre treatment of the dialysis tubing was to first boil for 15 minutes in 5% NaHCO3 and 0.01M EDTA, and then rinse with distilled water to remove diffusible salts and UV absorbing material (*21*). The dialysis tubing was stored under refrigeration (5 °C) in 50% ethanol-water until needed, and then rinsed with copious amounts of distilled water prior to use.

Characterization of the BSA solutions. Several complementary methods were employed in the characterization of the BSA solutions. Listed below are only those methods that have a direct bearing on the interpretation of the data.

Osmometry. Osmotic pressure (π, in atmospheres) measurements were made with a Knauer Membrane Osmometer Type 01.00 connected to a Kipp and Zonen (Holland) type B D 40/04/05 recorder. The temperature was monitored with a telethermometer model 4310 (Yellow Springs Instrument Co., Inc.) to an accuracy of ± 1 °C. The osmometer was equilibrated with pH adjusted KCl solutions overnight at 25 °C. The membranes used in these experiments were B-19 S&S deacetylated acetyl membranes (Catalog number 2001-5, suitable for molecular weights greater than 20,000). The plotting scale was calibrated to 10 cm of water. After attaining a constant baseline 0.4 mL aliquots of a given BSA solution was injected very carefully into the inlet tube of the osmometer and the osmotic pressure was recorded in units of cubic centimeters of water. The solutions at a given pH and ionic strength were studied in order of increasing concentration of BSA. Data for which drifts in the baseline of greater than 1% were not retained. The cell was thoroughly rinsed with the pH adjusted KCl solution after each set of measurements. The osmotic pressure was fitted to the cubic equation,

$$\pi = \frac{RT}{M_{BSA}} c_{BSA} + b_\pi c_{BSA}^2 + c_\pi c_{BSA}^3 \tag{16}$$

where M_{BSA} is the molecular weight of the BSA, c_{BSA} is the weight concentration (g/L) of the BSA solutions, and b_π and c_π are least-square fitting parameters to the data. Representative results are summarized in Table III.

Viscometry. A Cannon 100 A 45 capillary-flow viscometer (Cannon Instrument, Box 16, State College,. PA.) with a flow time of 60 seconds for water was used in these experiments. The viscometer was submerged in a water bath with a temperature control to ± 0.1 °C using a Haake E 12 thermostatic control. The time flow between the two marks was measured by an electronic digital stopwatch (catalog number 14-648, Fisher Scientific Co.). Between each run the viscometer was cleaned and washed with distilled water, followed by lab-grade acetone and dried at 120 °C. The relative viscosity of the BSA solutions was calculated from the ratio of the flow times t_{BSA} and t_{KCl}, for the BSA and KCl solutions, respectively. The relative viscosity obtained in these studies did not deviate substantially from unity.

Light scattering measurements. Both quasi-elastic and total intensity (I_{tils}) light scattering measurements were carried out on the same instrument. The cylindrical cells, which had an inlet and outlet tube, were cleaned with hot nitric acid (50% v/v) and then thoroughly rinsed with distilled water prior to use in the light scattering experiments. In preparation for the experiment the cell was placed in a specially designed "dust box" with a gravity-flow filtration system. In this arrangement the solutions to be filtered were placed in a reservoir outside of the dust box, where the unsupported Millipore filter was located at the bottom of the reservoir.

Tygon tubing connected to the inlet arm of the cell and filter reservoir assured that air was eliminated in the filtration process. To extract the liquid from the cell in the rinsing process a movable tapered glass tube was inserted in the other arm of the cell and connected to tygon tubing that led into a three-way valve, where one of the two outlets was connected to an aspirator and the other to the outside air. We found that this somewhat elaborate method eliminated the introduction of unfiltered air into the scattering cell. The filters used in these experiments were 0.45 µm Millipore filters. The cells were first rinsed several times with filtered distilled water, followed by two (or more) rinses with the desired KCl solution, and then one rinse with the BSA solution. This rinsing protocol was then followed by filtration of the BSA solution to be used in the light scattering experiments. The I_{tils} measurements were made at 5° intervals over the range 30° < θ < 120° using a digital multimeter. The absolute excess intensities were obtained using spectroscopic grade benzene as a standard, and correcting for the solvent contribution. The q-independent "solution structure factor" was then calculated from the expression,

$$S_{11} = \frac{I_{BSA}([KCl]) - I_0([KCl])}{I_{3.80}(0.1M) - I_0(0.1M)} \tag{17}$$

where $I_{BSA}([KCl])$ is the intensity of scattered light normalized to the benzene reference and BSA concentration that was measured for the dimensionless BSA concentration [BSA]x(10^5 M^{-1}) (the numerical value is the subscript) at the specified KCl concentration and $I_0([KCl])$ is the *intercept* of the $I_{BSA}([KCl])$ versus [BSA] profile corresponding to the solvent scattering. These measurements of the excess scattering intensity were then normalized to the scattering of a 3.80x10^{-5} M solution of BSA in 0.1 M KCl.

QELS measurements were carried out on the same BSA samples as the I_{tils} measurements. The autocorrelation functions were obtained with a Langley-Ford autocorrelator. These functions were characterized as a single exponential decay function with a baseline. Data were obtained at several delay times and extrapolation of these values to zero delay time provided the value of $D_{app}(q)$. In all [KCl] solutions no angle dependence was observed for $D_{app}(q)$, hence the equivalence $D_{app}(q) = D_{app}(q=0) = D_m$. were assumed. Linear regression analysis of the data obtained at pH 5.0 (near the isoelectric point) at 25 °C in 0.1 M KCl was found to be

$$D_m (25\ ^{\circ}C) = 6.58 \times 10^{-7}(1 - 0.023c'_{BSA})\ cm^2 / s$$ where c'_{BSA} is the concentration in g/mL (20). This expression is comparable to the 20 °C study of Oh and Johnson (22) for their highly purified BSA at pH 4.7 in 0.12 M NaCl/acetic acid, where the concentration coefficient was reported to be - 0.0166 mL/g. The difference may be attributed to the slight differences in pH, T, and electrolyte concentration. Plots of D_{app} versus $1/2C_{add}$ for BSA concentrations in the range 1-3 g/L and at four values of the pH are shown in Figure 1.

Discussion

In order to understand the complex diffusion process of macroions in moderate to low ionic strength solvents, it is necessary to examine the theoretical predictions of over simplified mathematical models. According to the results given in Table I for the added electrolyte - single macroion system, the CM model predicts that $D_{app}(q)$ decreases as q increases, and that this dependence is strongest for moderately charged macroions in extremely low ionic strength solvents. In the case of *highly charged* macroions the apparent q dependence is suppressed because $D_{app}(q)$ is near its low salt *plateau value*. The situation is somewhat more complex in the two macroion - no added electrolyte system. We hesitate to ascribe a specific motion to these eigenvalues and the resulting values of $D_{app(q)}$ given in Table II. Our caution is based on the study of Pusey, Fijnaut, and Vrij (23) in which they examined a system of two interacting (neutral) polymers. Based on the relative changes of the eigenvectors they interpreted the faster "+" mode in terms of a "collective mode" and the slower "-" mode in terms of an "exchange mode" for these dissimilar polymers. These assignments are consistent with the labels given, without proof, to the observed fast and slow relaxation modes in QELS studies. However, the fact that the two polymers move in opposite directions for the "-" mode does not necessarily mean that the motion is *towards* each other, as implied by an "exchange" of particles, since they could also be moving *away* from each other.

Our reluctance to associate eigenvalues with specific molecular motions is further supported by the behavior of these eigenvalues. From the low charge, low concentration limit in Table II, root 2 is identified with polyion 1 and root 3 with polyion 2. It is evident from the data in Table II that the eigenvalue identified with smaller of the two macroions is more affected by the electrical properties of the solution. In the case of polyions of equal size but different charge, the polyion with the smaller charge is more affected by concentration changes than the higher charged polyion which appears to retain its neutral charge value of D_1. It is not obvious to us at this time what type of coupled motion gives rise to these behavioral patterns.

Attention is now directed to the summary of the BSA results given in Figure 2 and Table IV for KCl solutions at pH 7.4. As shown in Figure 2, a decrease in $2C_{add}$ results in an increase in D_{app} and the functional dependence is stronger for the larger charge (higher pH). This behavior is consistent with equations 15 and 16. It is concluded from the data in Table IV that D_{app} obtained in 0.1 M KCl contains no interparticle interactions since the ratio R_A (cf. Table IV caption) is independent of [BSA], where $D_{1,h}^0$ is defined as D_1^0 at this salt concentration.

The mutual diffusion coefficient is related to the osmotic susceptibility by,

$$D_m = \frac{1}{f_m}\left(\frac{\partial\pi}{\partial C_1}\right)_{\mu,T} \tag{18}$$

The ratio R_B/R_A in Table IV is therefore a measure of the ratio $f_m / f_{1,h}^0$. The trend evident in Table IV is that R_B/R_A decreases as [BSA] increases, which is opposite to what one would expect if hydrodynamic interactions alone were operative in this system. However, this trend is in the correct direction for electroviscous effects since there is an increase in the counterion concentration concomitant with an increase in [BSA]. The ratio R_C/R_A is equation to $H_{11}(q)$ in accordance with equation 12. Since hydrodynamic interactions effectively appear as an attractive interaction the observation that $H_{11}(q)$ is greater than unity for 0.001 M KCl indicates that electrical rather than hydrodynamic interparticle interactions affect the magnitude of the apparent friction factor.

Direct rather than dissipation interparticle interactions are dominant for [KCl] = 0.001M since R_A increases with an increase in [BSA]. It is noted, however, that the inequality $(1/RT)(\partial\pi/\partial C_1) < 1/S_{11}$. ($R_B < R_C$, where R_B was computed from equation 16 using a molecular weight $M_{BSA} = 68,000$) prevails for the 0.001M KCl solvent. Failure of the generally accepted equality of equation 13 can be explained in terms of the reduced <u>intrinsic</u> scattering power of the polyions in the low ionic strength solvent. Stigter (24) and Vrij and Overbeek (25) showed that the scattered light intensity for the multicomponent polyion system is decreased when compared with the expected two-component system result. This is because the index of refraction increment at <u>constant chemical potential</u> must be used <u>instead</u> of at constant concentration. Vrij and Overbeek (25) obtained the following ratio,

$$\left[\frac{\left(\partial\tilde{n}/\partial c_1'\right)_\mu}{\left(\partial\tilde{n}/\partial c_1'\right)_c}\right]^2 = \left[1 - \frac{Z_1\alpha}{2}\frac{M_2}{M_1}\frac{\left(\partial\tilde{n}/\partial c_2'\right)_c}{\left(\partial\tilde{n}/\partial c_1'\right)_c}\right]^2 \tag{19}$$

where the subscript 1 (2) denotes the polyion (electrolyte), the subscript c' denotes constant concentrations of all other components, and the factor $0 < \alpha < 0.5$ is the

Table III. Least-Squares Fit of the Osmotic Pressure for BSA

pH	[KCl] (M)	$b_\pi \times 10^5$ (atm L^2 g^{-2})	$c_\pi \times 10^7$ (atm L^3 g^{-3})	$\sigma^{1/2} \times 10^4$ [a]
5.00	0.100	-0.0342	0.290	5.45
5.00	0.005	-0.277	0.847	3.21
6.50	0.100	-0.193	1.90	1.34
6.50	0.001	2.01	8.88	10.1
7.40	0.100	2.68	-15.3	2.64
7.40	0.010	2.68	3.59	1.08
7.40	0.0001	6.56	-14.2	1.80
8.50	0.100	-0.0129	2.86	2.06
8.50	0.005	3.72	-2.74	4.48
8.50	0.001	7.05	-6.43	4.64

$$a - \sigma^{1/2} = \left[\sum_{i=1}^{n} \left(x_{calculated} - x_{experimental} \right)^2 / n \right]^{1/2}$$

Table IV. Characterization of BSA in KCl at pH 7.4 and 25 °C

[KCl] (M)	c_{BSA} (g/L)	R_A	R_B	R_C	R_B/R_A	R_C/R_A
0.100	3.80	1.01	1.31	1.00	1.30	0.99
0.100	16.4	1.06	1.06	0.88	1.00	0.83
0.100	18.0	0.99	0.88	0.85	0.83	0.86
0.001	7.56	1.56	2.76	3.75	1.77	2.40
0.001	12.7	2.07	3.31	3.81	1.60	1.84
0.001	14.9	2.56	3.53	5.42	1.38	2.12

$$R_A = \frac{D_m}{D_{I,h}^0} ; \qquad R_B = \frac{1}{RT} \left(\frac{\partial \pi}{\partial [BSA]} \right)_{\mu,T} ; \qquad R_C = \frac{I_{3.80}(0.1M) - I_0(0.1M)}{I_{BSA}([KCl]) - I_0([KCl])}$$

deficit of coions, where the upper limit of 0.5 results for low potentials. Vrij and Overbeek (25) reported in their Table II the index of refraction increments of poly(methacrylic acid) at constant μ and constant c' in different salt solutions. Using their values one has $[(\partial\tilde{n}/\partial c_1)_\mu/(\partial\tilde{n}/\partial c_1)_{c'}]^2 = (0.209/0.234)^2 = 0.80$ in 0.1 M NaCl and $(0.169/0.227)^2 = 0.55$ in 0.01 M $(NH_4)_6Mo_7O_{24}$. Hence it is important to correct for the intrinsic scattering power of the macroions in the lower ionic strength solvents when using equation 13. An alternative explanation, however, is found by comparison of equations 12 and 15. Given that hydrodynamic interaction are negligible in comparison to the electrical effects, one has from equations 12 and 15 the difference $R_C - R_B = A_d Z_1^2 C_1/2C_s$. Using the two extreme BSA concentrations one obtains $(R_C - R_B)/c_{BSA} = A_d Z_1^2/2C_s M_{BSA} \approx 0.13$, from which one obtains for $A_d = 1$ the approximate value $Z_1 \approx -4$. In view of the approximations employed in the calculation, this value is in reasonable agreement with the value of $Z_1 = -9$ reported by Tanford *et al.* (26) for BSA in pH 7.4 buffer in 10^{-3} M electrolyte.

We now turn to the data of Petsev and Denkov (7) of PLS over the ranges $1\times10^{-4} < \phi_1 < 6\times10^{-4}$ and $10^{-4}M < [NaCl] < 10^{-2}M$. A detailed analysis of these data will be presented elsewhere. Of relevance to the present communication is that D_m (cf. equation 9) is a linear function of ϕ_1. The reported slopes, λ, are plotted as a function of $1/2C_{add}$ in Figure 2. The counterions from the macroions are not included since [PLS] is on the order of 5×10^{-8} M. Although there are only four points, these data indicate two linear regions. The solid line in Figure 2 is the equation

$$\lambda = \frac{0.173}{2C_{add}} + 135 \tag{20}$$

To estimate the charge from equation 20 we use the reported micrograph value of $R_S = 95\times10^{-8}$ cm (7) and assume $\delta_{11} + A_d = 2$, with the result $Z_1 \approx 14$. This is an extremely low estimate but is consistent with the very low results obtained for polyions using conventional polyelectrolyte models. The intercept, however, is much larger than the *maximum* prediction of the neutral hard sphere, viz., $8 - k_h$. It is tempting to suggest that this provides support for the existence of an "effective" hard sphere with a radius determined by the screening length $1/\kappa$, as proposed recently by Okubo (27). We do not, however, ascribe to this interpretation since it leads to unphysical results such as an erroneous value of Avagoadro's constant (28).

Conclusions

The reciprocal intensity of scattered light first normalized to theta conditions (high salt for polyions) is sometimes employed for the osmotic susceptibility. It is concluded from the BSA studies that this equivalence is valid only if the intensity is measured under low electrolyte concentration is first corrected for multicomponent solution effects. A second interpretation that deserves consideration is the "cell model" that explicitly separates small ion-macroion and macroion-macroion contributions. Although to be discussed in detail elsewhere, the "composite" diffusion coefficient appears to be adequate in the interpretation of the PLS data of Petsev and Denkov (7). It is quite clear that many effects must be taken into consideration when attempting to make sense • polyelectrolyte data in solutions with very low electrolyte concentrations.

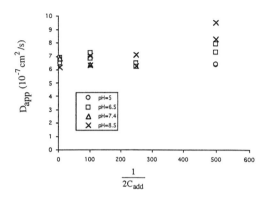

Figure 1. Electrolyte Concentration Dependence of D_{app} for BSA at Varying pH values. The BSA concentrations are in the range 1-3 g/L.

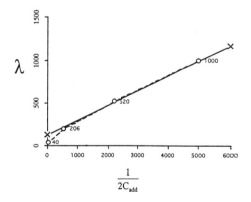

Figure 2. Electrolyte Concentration Dependence of Volume Fraction Coefficient. The data shown above were obtained from Petsev and Denkov (7). The solid line was computed from equation 20.

References

1. Giordano, R; Maisano. G.; Mallamace, F.; Micali, N.; Wanderlingh, F. *J. Chem. Phys.* **1981**, *74*, 2011-2015.
2. Carter, J. M.; Phillies, G. D. J. *J. Phys. Chem.* **1985**, *89*, 5118-5124.
3. Anderson, J. L.; Reed, C. C. *J. Chem. Phys.* **1976**, *64*, 3240-3250.
4. Lin, S.-C.; Lee, W. I.; Schurr, J. M. *Biopolymers* **1978**, *17.*, 1041-1064.
5. Schmitz, K. S. *An Introduction to Dynamic Light Scattering by Macromolecules*; Academic Press, Boston, MA., **1990** pp 207-210.
6. Tivant, P; Turq, P.; Drifford, M.; Magdelenat, H.; Menez, R. *Biopolymers* **1983**, *22*, 643-662.
7. Petsev, D. N.; Denkov, N. D. *J. Colloid and Interf. Sci.* **1992**, *149*, 329-344
8. Schmitz, K. S. *Macroions in Solution and Colloidal Suspension*, VCH Publishers, New York., N.Y., **1993** pp 140-145.
9. Belloni, L.; Drifford, M.; Turq, P. *J. Physique Lett.*, **1985**, *46*, L207-L215
10. Schurr, J. M. *Chem. Physics*, **1987**, *111*, 55-86.
11. Schurr, J. M., *Chem. Physics*, **1980**, *45*, 119-132.
12. Medina-Noyola, M.; Vizcarra-Rendón, A. *Phys. Rev. A* **1985**, *32*, 3596-3605.
13. Ruín-Estrada, H.; Vizcarra-Rendón, A.; Medina-Noyola, M.; Klein, R. *Phys. Rev. A* **1986**, *34*, 3446-3451.
14. Vizcarra-Rendón, A.; Ruín-Estrada, H.; Medina-Noyola, M.; Klein, R. *J. Chem. Phys.*, **1987**, *86*, 2976-2985.
15. Booth, F. *J. Chem. Phys.* **1954**, *22*, 1956-1968.
16. Geigenmüller, U. *Chem. Phys. Lett.* **1984**, *110*, 666-667.
17. Carter, J. M.; Phillies, G. D. J. *J. Phys. Chem.* **1985**, *89*, 5118-5124.
18. Imai, N.; Mandel, M. *Macromolecules* **1982**, *15*, 1562-1566.
19. Penfold, R.; Nordholm, S.; Jönsson, B. *J. Chem. Phys.* **1990**, *92*, 1915-1922.
20. Singh, N. *On the Possible Role of Small Ions in the Dynamics of Polyelectrolytes* **1986**, University of Missouri Ph. D. Dissertation.
21. Fu, J. C. C.; Gruenwedel, D. W. *Biopolymers* **1976**, *15*, 265-282.
22. Oh, Y. S.; Johnson, C. S. *J. Chem. Phys.* **1981**, *74*, 2717-2720.
23. Pusey, P. N.; Fijnaut, H. M.; Vrij, A. *J. Chem. Phys.* **1982**, *77*, 4270-4281.
24. Stigter, D. *J. Chem. Phys.* **1960**, *64*, 842-846.
25. Vrij, A.; Overbeek, J. Th. G. *J. Colloid Sci.* **1962**, *17*, 570-588.
26. Tanford, C.; Swanson, S. A.; Shore, W. S., *J. Am. Chem. Soc.* **1955**, *77*, 6414-6421.
27. Okubo, T. *Acc. Chem. Res.* **1988**, *21*, 281-286.
28. Ito, K.; Ieki, T.; Ise, N. *Langmuir*, **1992**, *8*, 2952-2956.

RECEIVED August 6, 1993

SYNTHESIS AND CHARACTERIZATION

Chapter 9

Reactive Polymers

Water-Soluble Vinyl-Terminated Oligomeric Poly(β-alanine)

Sun-Yi Huang and M. M. Fisher[1]

Stamford Laboratory, Cytec Industries, 1937 West Main Street, Stamford, CT 06904-0060

Low molecular weight water soluble vinyl terminated oligomeric poly (beta-alanine) was synthesized using organolithium initiators. Vinyl terminated poly (beta-alanine) consists of N-2-carboxyamidoethyl-acrylamide and oligomers which were fully characterized by analytical methods. The results show that one vinyl terminated group per chain was produced. Reaction time and temperature were the main variables used to control the N-2-carboxyamidoethylacrylamide content, molecular weight, and molecular weight distribution. N-2-carboxyamidoethyl-acrylamide was homopolymerized in aqueous medium by redox initiation. Poly (N-2-carboxyamidoethylacrylamide) was converted to a Mannich derivative and was quaternized. This new cationic water soluble polymer is an effective organic flocculant for bentonite clarification.

Block and graft copolymers represent an important class of materials which are achieving ever increasing commercial success. Examples are Shell's Kraton, Phillips' Salprene, DuPont's Hytrel, B. F. Goodrich's Estines and Cyanamid's XT polymers. The field of block and graft copolymers has been extensively reviewed in a number of recent monographs (1-3). A significant development in the area of graft copolymer synthesis is the concept of copolymerizing conventional monomers with vinyl terminated prepolymers (4-6). The principal advantage of this approach is the avoidance of a homopolymer which is always obtained when conventional free radical grafting techniques are used. The potential application of vinyl terminated prepolymers, in a number of areas of commercial significance to industries has served as the basis for the development of a novel graft copolymer useful as paper additives, flocculants, fibers, and thermoplastic elastomers. The general purpose was to improve the properties of these materials by incorporating graft segments capable of strong side chain internations. It was

[1]Current address: American Plastic Council, 1275 K Street, NW, Washington, DC 20005

NOTE: This chapter is Part III in a series of articles.

envisioned that a macromonomer with desired properties could be synthesized through the base initiated hydrogen transfer polymerization of acrylamide (7):

$$B^{\ominus} + nCH_2 = CH \longrightarrow CH_2 = CH - \underset{\underset{O}{\parallel}}{C} - NH - (CH_2 - CH_2 - \underset{\underset{O}{\parallel}}{C} - NH-)_{n-1}H \quad (I)$$
$$\underset{\underset{NH_2}{|}}{\underset{C=O}{|}}$$

In 1954, a patent by Matlack (7), and papers by Breslow, et al. (8,9) described the anionic polymerization of acrylamide to high molecular weight poly (beta-alanine). Breslow, et al. suggested two possible mechanisms for the base initiated polymerization of acrylamide. One is a proton abstraction from the amide group by the base in reaction (II).

$$B^{\ominus} + CH_2 = CH - \overset{\overset{O}{\parallel}}{C} - NH_2 \longrightarrow CH_2 = \overset{\overset{O}{\parallel}}{C}NH^{\ominus} + BH \quad (II)$$

The second is the Michael addition of the initiator to the carbon carbon double bond of acrylamide in reaction (III).

$$B^{\ominus} + CH_2 = CH - \overset{\overset{O}{\parallel}}{C} - NH_2 \longrightarrow B - CH_2 - \overset{\ominus}{C}H - \overset{\overset{O}{\parallel}}{C}NH_2 \longrightarrow$$
$$B - CH_2 - CH_2\overset{\overset{O}{\parallel}}{C}NH^{\ominus} \quad (III)$$

They concluded that the predominant initiation reaction is (II) because a solid material from a reaction mixture of potassium tert-butoxide and acrylamide was a dimer or an oligomer containing a vinyl group. Ogata (10) concluded that nucleophilic addition to the vinyl group must be the initiation step for the polymerization of acrylamide using sodium methoxide by infrared spectroscopy. Tani, et al. (11) studied the products of the reaction of acrylamide with sodium methoxide and sodium tert-butoxide. They found that the tert-butoxide anion abstracts a proton from amide groups and adds to the carbon-carbon double bond, but the methoxide anion only adds to the carbon-carbon double bond. Trossarelli, et al. (12) found that the amounts of tert-butanol in the reaction products agreed with those calculated from the quantity of sodium tert-butoxide employed. Leoni, et al. (13) isolated dimeric and trimeric products from alkyl lithium initiated polymerization of acrylamide. Both groups supported the initiation mechanism in equation (II). Kobayashi, et al. (14) observed that acrylamide was readily polymerized to poly (beta-alanine) by sodium cellulosate and sodium polyvinyl alcohol initiators. No block and graft copolymers were found. However, the reaction led to carbamoylethylation of the polymer backbones which supported the initiation mechanism of equation (II). Moore, et al. (15,16) found that in-situ polymerization of acrylamide initiated by poly (phenylene terephthalamide) anion in solution in vacuo at 80-115°C gave molecular composites of a poly (beta-alanine)-graft copolymer and poly (beta-alanine) homopolymer. The result supported the reactions (II) and (III). It is very difficult to draw conclusions based on the conflicting evidence reviewed above. Initiation of the hydrogen transfer polymerization of acrylamide appears to be dependent on the nature of the bases. Many other factors such as reaction temperature and medium are also expected to influence the reaction. It was also found that the basicity of the initiator and steric effects influence the mode of initiation and the ultimate degree of unsaturation of the poly (beta-alanine) (17). Tarvin (18) successfully copolymerized a high vinyl content of poly (beta-alanine) with acrylonitrile and methyl methacrylate. The terpolymers have wet fusion points at 175°C.

In this paper, we have focused on the synthesis and characterization of low molecular weight vinyl terminated poly (beta-alanine) (VTN-3). To obtain a better understanding of the polymerization process, acrylamide was converted to the low molecular weight VTN-3 and N-2-carboxyamidoethylacrylamide (AMD dimer) in various solvents and the rate of formation was measured. We report and discuss the molecular weight determination obtained by gel-permeation chromatographic separation of oligomer and AMD dimer, vapor phase osmometry, NMR, and bromine-bromate titration for unsaturation. AMD dimer was successfully isolated from oligomeric mixtures. Homopolymerization of this monomer produced a high molecular weight water soluble polymer using an aqueous redox system. The preparation of Mannich derivatives and subsequent quaternization with dimethylsulfate was investigated for the cationic derivatives of poly (acrylamide dimer) covering a wide range of molecular weight. This new cationic water soluble polymer has been studied as an effective organic flocculant for bentonite water clarification.

Experimental

VTN-3 Oligomer Preparation. Ten g of sublimed acrylamide and 100 ml of p-dioxane distilled from CaH_2 are placed in a 250 ml, 4-neck, round-bottom flask equipped with a stirrer, N_2 inlet, thermometer, condenser, and rubber septum. The mixture is stirred in a N_2 atmosphere at ambient temperature for 30 mins. until all of the acrylamide is dissolved. The acrylamide solution is thermostated at 25°C, vigorously stirred, and 4.3 ml (1.58 M) of n-butyl lithium solution slowly added through the rubber septum by a syringe. The reaction between acrylamide and n-butyl lithium is exothermic and the addition is carried out slowly over a few mins. to maintain at 25°C. After 24 hrs. 1 ml of water is added to terminate the reaction. The white dispersed solid is separated from p-dioxane by centrifugation. The solid phase is dried under vacuum at 60°C for 24 hrs. The yield is 9 g of oligomer.

Synthesis of Acrylamide Dimer [N-2-Carboxyamidoethylamide]. AMD dimer was obtained in 75% yield by triple ultrafiltration of a crude VTN-3 reaction mixture prepared in p-dioxane at ambient temperature. A 2-3% aqueous solution of the initial reaction product was pressurized through an Amicon Model 402 stirred ultrafiltration cell equipped with a UM-5 membrane. The filtrate purity was 98-100% of AMD dimer. An alternative procedure for obtaining pure dimer in 40% yield was to heat the same crude VTN-3 product in p-dioxane to reflux. The slurry was quickly filtered to separate the insoluble oligomer. The dimer recrystallized from the filtrate on cooling.

Homopolymerization of AMD Dimer. Polymerization of AMD dimer was carried out in a 250 ml 3-neck flask containing 4 g of AMD dimer and 96 g deionized water at pH=6.0. The initiation temperature was 40°C. Sodium metabisulfite concentration was varied between 38-5000 ppm based on monomer while the ammonium persulfate concentration was 1500-30000 ppm based on monomer. The polymerization was complete in six hrs. In some runs, 4,4'-azobis (4-cyanovaleric acid) was used as the initiator.

Mannich and Quaterization on Poly(acrylamide dimer). The preparation of Mannich derivatives of poly (acrylamide dimer) using formaldehyde and dimethylamine and subsequent quaternization with dimethyl sulfate were carried out as follows: A solution of freeze-dried polymer, 0.5 g in 10.22 g of water was prepared. 0.91 g of 35% $(CH_3)_2NH$ solution was added followed by 0.48 g of 44% HCHO solution. The reaction was carried out at ambient temperature at pH=11 for 6 hrs. and then 0.89 g of $(CH_3)_2SO_4$ was slowly added. After 12 hrs. the Mannich quaternary polymer solution was analyzed (19) and cationic equivalent (CEQ) was found to be 5.14 meq/g. The degree of substitution is defined as [CEQ/3.08 - 1]x100% where 3.08 is the theoretical CEQ of 100% quaternization. A 67% quaternization was achieved. The quaternary polymer was dialyzed for 2 days in Spectrapor membrane (m.wt. cut off 2,000) and freeze-dried. The CEQ on the dialyzed product was 2.15.

Kinetic Measurement. Aliquots of the reaction mixture were removed at various times and analyzed for the residual AMD monomer using a Hewlett Packard 7620 Research Gas Chromatograph. The column was packed with 8'x20 mm 20% Carbowax 20M WA-N-DMCS 60/80 mesh and helium was used as the carrier gas.

Characterization. Number average molecular weights were determined by a Knaner Dampf-Druck vapor phase osmometer. GPC analyses were performed using a Perkin-Elmer M601 equipped with 10^6, 10^5, 10^4, 10^3 Å styragel columns. Columns were standarized by purified standard VTN-3 samples, Mw = 141, 13,800, 19,200, 20,200, 28,200; lnMi = a + bVi, where a = $1.545.10^8$, and b = 36.2. Samples concentrations were 0.1 - 0.5% by weight in hexafluoro-isopropanol (0.01 M KOAC). GPC analyses were also performed using Sephadex G-10 and G-25. G-10 and G-25 were packed with cross-linked polysaccharide swollen in 0.1M $NaNO_3$ solution.

Flocculation of Bentonite Suspensions. Bentonite flocculation results for the cationic poly (AMD dimer) samples and commercial cationic polyamine and poly (diallydimethylammonium chloride) were compared. The standard practice (20) was followed where the dosage, in ppm, of real polymer required to reduce the slurry transmittance to 20% of the initial value, T_{20}, was used as a measure of the relative flocculation efficiency.

Results and Discussion

Synthesis of VTN-3 Oligomer. The high temperature polymerization of acrylamide in pyridine using t-butyl lithium initiator was discussed (*17*) and it is the preferred method for preparing water soluble poly (beta-alanine) in the 10,000 -20,000 molecular weight range. The polymer was shown to contain an average of one vinyl group per chain. A synthetic route to obtain lower molecular weight poly (beta-alanine) in 141 - 1,000 molecular weight range was desired in order to have available a wide range of macromonomer for copolymerization studies. A process using p-dioxane as the solvent, alkyl lithium initiators and a reaction temperature of ~25°C was developed. The major reaction product is an oligomer which corresponds to the structure of $CH_2=CHCONH(CH_2CH_2CONH)_nH$, where n = 1 - 10. Reaction conditions and product characteristics are summarized in (Table I). Since the degree of polymerization under these conditions is low, NMR, VPO, and unsaturation determination could be used for detailed structural analysis. In particular, integrated NMR spectra were used to determine number average molecular weights. These values reported in column 11 of Table I are in good agreement with $\overline{Mn}$ values calculated from the degree of unsaturation and VPO data (± 10-15%). In contrast to high temperature polymerizations where only t-butyl lithium gave high degree of vinyl termination, all alkyl lithium compounds examined in this series of experiments produced high yields of vinyl terminated oligomer. Apparently proton abstraction from the amide group is favored at low temperature even for unhindered alkyl lithiums. These oligomers are highly water soluble and soluble in DMSO. They are partially soluble in absolute alcohol, methanol but insoluble in non-polar organic solvents. DSC analysis indicated decomposition above 210°C rather than true melting. The proposed mechanism of acrylamide polymerization with organolithium compounds is as follows:

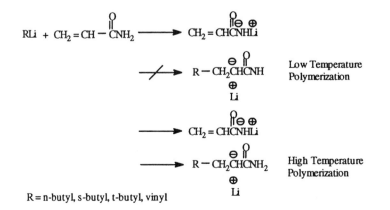

R = n-butyl, s-butyl, t-butyl, vinyl

INITIATION.

$$CH_2 = CHCNHLi + CH_2 = CHCNH_2 \longrightarrow CH_2 = CHCNHCH_2CHCNH_2 \longrightarrow$$

$$CH_2 = CHCNHCH_2CH_2CNHLi$$

PROPAGATION

$$CH_2 = CHCNHCH_2CH_2CNHLi + CH_2 = CHCNH_2 \longrightarrow$$

$$CH_2 = CHCNHCH_2CH_2CNH \sim\sim\sim CH_2CH_2CNH\, Li$$

TERMINATION.

$$CH_2 = CHCNHCH_2CH_2CNHLi + CH_2=CHCNH_2 \longrightarrow$$

$$CH_2 = CHCNHCH_2CH_2CNH_2 \downarrow \quad \text{in p-dioxane}$$
$$+$$
$$CH_2 = CHCNH\, Li$$

$$CH_2 = CHCNHCH_2CH_2CNHLi + CH_2=CHCNH_2 \xrightarrow{\text{DMF or} \atop \text{DMSO}}$$

$$CH_2 = CHCNHCH_2CH_2CNH \sim\sim\sim CH_2CH_2CNH_2 + CH_2 = CHCNH\, Li$$

CHAIN TRANSFER.

$$CH_2 = CHCNHCH_2CH_2CNH \sim\sim\sim CH_2CH_2CNH\, Li + CH_2 = CHCNH_2 \longrightarrow$$

$$CH_2 = CHCNHCH_2CH_2CNH \sim\sim\sim CH_2CH_2CNH_2 + CH_2 = CHCNH\, Li$$

Table I. Synthesis of Vinyl Terminated Poly (beta-alanine) Oligomer

Sample No.	Initiator	Monomer Initiator (mole/mole)	Temp. (°C)	Time (hr)	Conversion (%)	ηinh^a	$\bar{M}_n^b$ by Unsaturation	$\bar{M}_n^c$	$\bar{M}_n$ by NMR	Unsaturation %
118-2	n-Butyl	21.3	25	24	75	0.040	286		270	106
118-1	sec-Butyl Li	20.9	25	24	75	0.039	289	290	256	113
119-1	t-Butyl Li	20.4	25	22	90	0.032	260	270	231	112
119-3	t-Butyl Li	25.5	25	24	79	----	---		256	---
119-2	t-Butyl Li	20.1	25	40	85	0.046	252		276	91
117-C	Vinyl Li	25.2	25	24	83	0.030	---		234	---
117-B	Vinyl Li	25.2	25	24	90	0.039	430	430	490	88

[a] Inherent viscosities were measured for 1% polymer in 1N aqueous NaCl.
[b] Unsaturation was determined by bromine-bromate titration.
[c] $\bar{M}_n$ was determined by vapor phase osmometry.

Effect of Temperature on the Conversion of Acrylamide to AMD Dimer and Oligomer. For organolithium initiated oligomerization of acrylamide in p-dioxane, the relative yield of dimer and oligomer was found to depend strongly on the reaction temperature. Results are shown in (Table II). A large amount of acrylamide dimer is produced in the early stages of the polymerization especially when the lowest polymerization temperatures were used. Therefore, pure dimer is isolated essentially free from oligomer contamination. The concentration of dimer linearly decreased from 80% to 20% as the polymerization temperature was increased from 16°C to 50°C. Correspondingly, the concentration of oligomer linearly increased from 15% to 70%. The overall conversion was ~90%. The reason that dimer can readily be isolated is that acrylamide dimer has a very low solubility in p-dioxane (*13*). The formation of oligomer occurs predominantly after dimer has precipitated. The molecular weight distribution of oligomer was also found to depend on reaction temperature. $\overline{M}w/\overline{M}n$ increases as the reaction temperature increases. For reaction temperatures between 15°C and 25°C, $\overline{M}w/\overline{M}n$ is about 2 - 3 in an agreement with a step-growth mechanism. However, when the reaction temperature is the range of 30 - 50°C, $\overline{M}w/\overline{M}n > 3$ which may indicate chain branching is occurring. The formation of acrylamide dimer was also shown to depend on the nature of the solvent. DMF or DMSO are solvents for AMD dimer and oligomerization past the dimer stage occurs readily as shown in (Figure 1). (Figures 2 and 3) are typical GPC curves of VTN-3.

Solvent Effect. Both the rate of reaction and formation of oligomers strongly depend on solvent. Comparable reactions were carried out in DMSO, DMF, and p-dioxane. t-Butyl Li was used as the initiator under identical reaction conditions in three different solvents. (Figure 4) shows that 80% of the acrylamide monomer has disappeared in 30 minutes in DMSO. In the same experiment, 65% of the monomer has disappeared in DMF. Only 50% of the monomer is converted to oligomer in p-dioxane. After ~1 hour of reaction time, acrylamide dimer starts to precipitate in p-dioxane but remains in solution in DMF and DMSO. A homogeneous phase is obtained throughout the reaction in DMF and DMSO solvents and a heterogeneous phase in p-dioxane.

Poly (acrylamide dimer). Homopolymerization of acrylamide dimer produced a relatively low molecular weight polymer. Since higher molecular weights were desired for paper additives and water treating applications, an effort was made to upgrade the yield and molecular weight of this polymer. Initiator types, initiator concentration, temperature, and the reaction time were varied for a series of acrylamide dimer polymerizations in aqueous medium, and the results are summaried in (Table III). Initiator levels of ~15,000 ppm persulfate and 35-75 ppm of metabisulfite resulted in high conversion and acceptable viscosity. Yields dropped off rapidly below 15,000 ppm persulfate and high conversion but low viscosities were obtained when the metabisulfite concentration was increased in contrast to acrylamide. Acrylamide (sample no. 101-4) could be polymerized to a very high molecular weight polymer under similar conditions. This bulk viscosity is about the weight-average molecular weight of greater than 10^6 of polyacrylamide (*21*).

Table II. The effect of Temperature on the Acrylamide Dimer Content and the Molecular Weight[a]

Sample No.	Initiator	Temp. °C	Time (hrs.)	Conv. (%)	AM Dimer[b] Content (%)	Oligomer[c] Content (%)	$\bar{M}_w$	$\bar{M}_n$	$\bar{M}_w/\bar{M}_w$	ηinh[d]
58-1	n-BuLi	10	16	98	80	18	-----	-----	----	0.02
5-1		15	24	85	57	28	7,910	2,900	2.73	0.04
5-5		20	2	50	50	1	-----	-----	----	---
5-4		20	24	92	55	37	11,000	3,600	2.99	0.03
5-2		31	24	85	33	52	14,000	4,100	3.39	0.06
5-3		41	24	90	13	76	29,400	5,880	4.49	0.08
6-3	n-BuLi	15	24	85	57	28	6,200	2,600	2.38	0.04
6-1		18	24	83	51	32	3,750	11,000	2.64	0.05
6-4		35	83	83	17	66	22,000	5,600	3.96	0.06
6-2		50	24	85	13	72	21,000	5,970	3.51	0.08
6-5	n-BuLi	16	24	85	71	14	-----	-----	----	0.03
6-6		25	24	90	52	38	-----	-----	----	0.05
6-7		35	24	85	36	49	-----	-----	----	0.06
6-8		45	24	90	21	69	-----	-----	----	0.07

[a] Reaction conditions: 1). 20g acrylamide, 200 ml p-dioxane, 8.8 ml (1.58M) n-BuLi. 2). 20g acrylamide, 200 ml p-dioxane, 9.5 ml (1.5M) s-BuLi. 3). 20g acrylamide, 200 ml p-dioxane, 6.3 ml (2.30M) t-BuLi.
[b] Acrylamide dimer and oligomer contents calculated from GPC data.
[c] Oligomer molecular weight distribution was calculated from integrated GPC chromatograms.
[d] Inherent viscosities were determined for 1% polymer in 1N NaC1 aqueous solution.

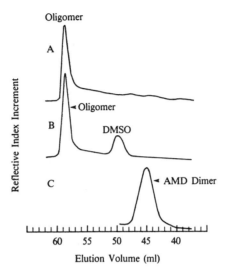

Figure 1. GPC chromatogram of vinyl terminated poly (beta-alanine) oligomer prepared in DMF and DMSO. A. 20 g acrylamide, 200 ml DMF, 6.4 ml t-BuLi (1.8 M). B. 20 g acrylamide, 200 ml DMSO, 6.4 ml t-Buli (1.8 M). C. acrylamide dimer.

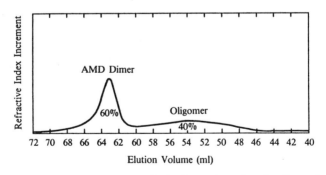

Figure 2. GPC chromatogram of vinyl terminated poly (beta-alanine) oligomer (Perkin-Elmer Styragel Column).

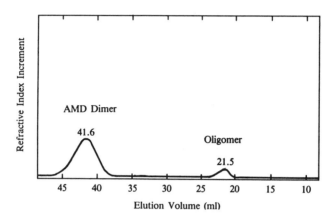

Figure 3. GPC chromatogram of vinyl terminated poly (beta-alanine) oligomer (Sephadex G-10 Column).

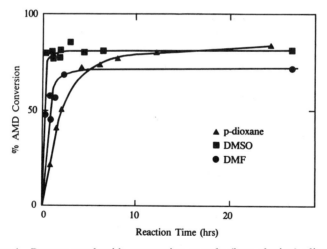

Figure 4. Percent acrylamide conversion to poly (beta-alanine) oligomer as function of time using t-Buli as initiator in different solvents: ■ DMSO ● DMF ▲ p-dioxane.

Table III. Homopolymerization of N-2-Carboxyamidoethylacrylamide in Aqueous Solution at pH 6.5-7

Sample No.	$(NH_4)_2S_2O_8$ ppm[a]	$Na_2S_2O_5$ ppm[a]	ADVA[b] ppm[a]	Monomer Conc.,%	Temp. °C	Time Hrs.	Residual AM Dimer,%	BV[d] (cps)
82-2A	30,000	5,000	---	4	40	6	T[c]	2.25
82-2B	20,000	5,000	---	4	40	6	T	2.10
83-2B	20,000	5,000	---	4	50	6	T	2.10
83-2C	15,000	1,000	---	4	40	6	11	---
84-1A	10,000	175	---	4	40	6	45	---
100-1	15,000	37.5	---	4	40	6	T	620[e]
100-2	15,000	75	---	4	40	6	T	575
101-1	1,500	37.5	---	4	40	6	L[c]	---
101-4	15,000	37.5	---	4	40	1	10	10,000
				(acrylamide)			(acrylamide)	
55-4	---	---	6,300	9	50	6	16	high[f]
97-1	---	---	6,000	4	60	4.5		medium
101-3	---	---	1,000	4	50	6		180

[a] ppm are based on monomer weight.
[b] ADVA is 4,4'-azobis(4-cyanovaleric acid).
[c] T(trace), L(large). Residual AM dimer determined using GPC.
[d] Bulk viscosity.
[e] ηinh (0.1% polymer in 1M NaCl) = 3.82 after dialyzing and freeze drying.
[f] ηinh = 5.0.

A water soluble azo initiator, 4,4'-azobis (4-cyanovaleric acid) (ADVA) was also used in some experiments. Molecular weight decreased with decreasing ADVA concentration. Additional experiments confirmed that higher molecular weights could be achieved with ADVA than with the aqueous redox system.

Mannich Quaternary of Poly (acrylamide dimer). Poly (acrylamide dimer) after conversion to the Mannich derivative and quaternization, contained cationic functions separated from the polymer backbone by an 8-atom spacer group:

$$CH \! - \! C \! - \! NH \! - \! CH_2 \! - \! CH_2 \! - \! C \! - \! NH \! - \! CH_2 \! - \! \overset{\oplus}{N}(CH_3)_3 \ \ CH_3(SO_4)^{\ominus}$$

The preparation of Mannich derivatives and subsequent quaternization with dimethyl sulfate was investigated for poly (acrylamide dimer) covering a range of molecular weights.

Reaction conditions and cationic equivalent data are shown in (Table IV). Mole ratios of polymer, formaldehyde, dimethylamine and dimethylsulfate were 1:1:1:1, 1:1:2:2 and 1:2:2:2. The degree of quaternization increased with an increase in the formaldehyde/polymer ratio, but increasing the DMA:HCHO ratio had little effect. Cationic equivalent data per unit weight of cationic poly (AMD dimer) (non-isolated samples) and per unit weight of cationic poly (AMD dimer) (isolated after dialysis) indicated that a cationcity of at least 70% was achieved.

Flocculation of Bentonite Suspensions Using Cationic Poly (acrylamide dimer). Bentonite flocculation results for the poly (AMD dimer) samples and commercial cationic polyamine and poly (diallyldimethylammonium chloride) are

Table IV. Preparation of Cationic Poly (Acrylamide Dimer)

Sample No.	Polymer (g)	ηinh[a]	H_2O (g)	HCHO (g)	$(CH_3)_2NH$ (g)	$(CH_3)_2SO_4$ (g)	CEQ[b]	CEQ[c]	DQ[d]
98-1	0.93	2.23	19.42	0.20	0.59	1.68	3.48		13
98-2	0.96	2.50	20.13	0.20	0.61	1.71	3.36		9
103-1	0.50	3.83	12.00	0.10	0.16	0.44	4.20		36
103-2	0.50	3.78	12.00	0.10	0.16	0.44	4.48		45
103-3	0.50	4.41	12.00	0.10	0.16	0.44	2.98		--
106-2	0.50	0.95	10.22	0.21	0.32	0.89	5.39		75
106-3	0.50	0.95	10.47	0.21	0.32	0.89	5.24		70
106-4	0.50	0.95	10.22	0.21	0.32	0.89	5.41	0.90	76
107-1	0.50	3.78	12.00	0.10	0.16	0.44	4.46	2.50	45
107-2	0.50	3.78	10.22	0.21	0.32	0.89	5.00	2.60	62
107-3	0.50	2.50	10.22	0.21	0.32	0.89	5.29	2.15	72
107-4	0.50	4.41	10.22	0.21	0.32	0.89	5.14		67

a. 0.1% polymer in aqueous 1M NaCl.
b. Cationic equivalent per gram of backbone polymer.
c. Cationic equivalent per gram of quaternary poly (acrylamide dimer). Quaternary polymers were isolated and freeze-dried after dialysis.
d. Degree of quaternization was defined as [CEQ/3.08 -1]x100% where 3.08 is the theoretical CEQ assuming 100% quaternization.

reported in (Table V). Standard practice uses the dosage in ppm of polymer solution required to reduce the slurry transmittance to 20% of its initial value, T_{20}, as a measure of the relative efficiency. Because the true weight percent of cationic poly (AMD dimer) could not readily be determined for each of the experimental polymers, T_{20} dosages could not be used to consistently evaluate relative efficiency. Since one of the objectives of this work was to determine flocculation efficiency inherent to the poly dimer structure, T_{20} and cationic equivalent data were combined in order to evaluate relative efficiency at constant cationicity. The results of this calculation are shown in column 5 in Table V. All of the experimental polymers were superior to polyamine on a molar charge basis. Efficiency tended to increase with decreasing molecular weight and increasing charge density. Lowest molecular weight samples 106-3, 106-4, 161-3 were superior to polyDADM. Limited stability tests showed that performance of cationic poly (AMD dimer) begins to decrease after two weeks at room temperature. However, the results of Series II demonstrate that cationic poly (AMD dimer) is generally less effective than the commercial polymers on an equivalent weight basis. An exception is the lowest molecular weight sample 161-3 which was found to be much more effective than any of the commercial polymers on a molar charge and equivalent weight basis.

The adsorbed charge distribution over the substrate surface by the low molecular weight cationic polymers should be more even than the charge distribution by high molecular weight chains. Therefore, loop dimensions of the adsorbed chain must be very small in comparison to the substrate hydrodynamic size. Consequently, the cationic polymer has folded down closer to the substrate particle surface. Such contacts will be prevented by the electrical repulsion between the substrate particles, thus flocculation would take place.

The flocculation of suspensions containing either bentonite clay or model polystyrene substrate, using commercially available cationic polyelectrolytes (22), showed that in most cases the dose of the flocculant was independent of its molecular weight. The charge density was the predominant factor in correlating with the optimum dose. However, the size and strength of a floc is dependent on the molecular weight of the polymer.

The effect of charge spacing between the quaternary nitrogens along the polymer chain on flocculation has not yet been studied. A wide range of molecular weights, and especially the very low molecular weight cationic polyelectrolytes on flocculation has not yet fully investigated. These will be explored as a future research direction.

Conclusions

Water soluble vinyl terminated oligomeric poly (beta-alanine) was synthesized with one vinyl group per chain. The reaction time and temperature were varied to control the molecular weight, molecular weight distribution, and AMD dimer content. AMD dimer and the oligomeric the VTN-3 were fully characterized by NMR, GPC, viscosity, vapor phase osmometry, and unsaturation determination. AMD dimer was homopolymerized in aqueous medium. Poly (AMD dimer) was successfully converted to the Mannich derivative and quaternary product. This

Table V. Flocculation of Bentonite Suspensions Using Cationic Poly (acrylamide Dimer)

Ref. No.	Dosage at T_{20} (ppm)	CEQ^a (meq/g)	ηinh^b	Efficiency $(\%)^c$
Series I				
98-1	2.25	3.48	2.23	141
98-2	2.83	3.36	2.50	116
103-1	2.05	4.20	3.83	128
103-3	3.00	2.98	4.41	126
Polyamine	1.75	6.30		100
Series II				
107-1-D	6.60	0.90	3.78	141
107-2-D	2.50	0.50	3.78	135
107-3-D	2.10	2.60	2.50	154
107-4-D	3.35	2.15	4.41	117
Polyamine	1.35	6.25		100
Series III				
161-3	0.50	5.40	0.50	296
106-2	0.82	5.39	0.95	180
107-1	1.30	4.46	3.78	138
107-2	0.95	5.00	3.78	168
107-4	1.22	5.14	4.41	128
Polyamine	1.28	6.25		100
Series IV				
106-3	0.90	5.24	0.94	167
106-4	0.90	5.40	0.95	159
Polyamine	1.24	6.21		100
Poly(DADM)	1.12	4.68		147

[a] CEQ values are based on unit weight of backbone polymer except for dialyzed samples in Series II.
[b] 1% polymer in aqueous 1M NaCl.
[c] Efficiency = [Dosage(T_{20})xCEQ]x100%/[Dosage(T_{20})$_r$xCEQ$_r$]; r is polyamine.

new cationic water soluble polyelectrolyte has been shown to be an effective organic flocculant for the clarification of bentonite.

Acknowledgments

The authors would like to thank Dr. J. Miller for viscosity, GPC and molecular weight measurements, Mr. Ray H. Anderson for unsaturation determination, Dr. J. Lancaster and Ms. M. Yao for NMR analysis, Dr. Fred Halverson and Ms. Marybeth Gutafason for flocculation studies, also Ms. Inez Minor and Ms. Valerie D. McNeil for the manuscript typing.

Literature Cited

1. Noshay, A.; McGrath, J. E. *Block Copolymers*; Academic Press: New York, 1977.
2. *Cationic Graft Copolymerization*; Kennedy, J. P. Ed.; *J. Appl. Polym. Symp.* No. 30; John Wiley & Sons: New York, 1977.
3. Mark, H. F. ; Overberger, C. G. *Encyclopedia of Polym. Sci. and Eng.*; John Wiley & Sons: New York, 1985; Vol. 2, p.324; 1986; Vol. 5, p.416.
4. *Macromer, Monomer Presentation*; GPC International Inc.: Englewood Cliffs, NJ, 1979.
5. Entelis, S. G.; Evreinov, V. V.; Gorshkov, A. V. *Adv. Polym. Sci.* 1985, Vol. 76, 129.
6. Kennedy, J. P.; Huang, Sun-Yi; Smith, R. R. *J. Macromol. Sci.* 1980, *A14(7)*, 1085.
7. Matlack, A. S. U. S. Pat. 2,672,480, 1954.
8. Breslow, D. S.; Hulse, G. E.; Matlack, A. S. *J. Am. Chem. Soc.* 1957, *79*, 3760.
9. Bush, L. W.; Breslow, D. S. *Macromolecules* **1968**, *1*, 189.
10. Ogata, N., *Makromol. Chem.* **1960**, *40*, 55.
11. Tani, H.; Oguni, N.; Akraki, I.; *Makromol. Chem.* **1964**, *76*, 82.
12. Trosarrelli, L.; Guaita, M.; Camino, G. *J. Polym. Sci., C.* **1969**, *22*, 721.
13. Leoni, A.; Franco, S.; Polla, G. *J. Polym. Sci.* **1968**, *A-1*, *6*, 3187.
14. Kobayashi, Y. *J. Polym. Sci., Polym. Lett. Ed.* **1976**, *14*, 299.
15. Moore, D. R.; Mathias, L. J. *Polym. Mater. Sci. Eng.* **1985**, *53*, 693.
16. Moore, D. R.; Mathias, L. J. *J. Appl. Polym. Sci.* **1986**, *32 (8)*, 6299.
17. Huang, Sun-Yi, *Polym. Prepr.* **1983**, *Vol. 24*, No. 2, 60.
18. Tarvin, R. F, Cytec Industries, personal communication.
19. Ueno, K.; Kina, Ken'yu, *J. Chem. Ed.* **1985**, *62 (7)*, 627.
20. Huang, Sun-Yi; Fisher, M. M. U. S. Pat. 4,247,432, 1981.
21. *Polyacrylamide, New Product Bull. Collective Vol. III*, No. 34; Am. Cyanamid Co.: Stamford, Ct., **1955**.
22. Dentel, S. K. Crit. *Rev. Environ. Control* **1991**, *21 (1)*, 41.

RECEIVED August 6, 1993

Chapter 10

Molecular-Weight Distributions of Water-Soluble Polyelectrolytes

Gel Permeation Chromatography Spectrophotometry—Low-Angle Laser Light Scattering

Ellen M. Meyer and Stephen R. Vasconcellos

Betz Industrial, 4636 Somerton Road, Trevose, PA 19053-6783

Recent developments in the use of gel permeation chromatography (GPC) for the determination of the molecular weight distributions of water soluble cationic polyelectrolytes are discussed. Absolute molecular weight measurements were made using a low-angle laser light scattering (LALLS) spectrophotometer. The polymers studied included five samples of an epichlorohydrin dimethylamine (epi/dma) copolymer with weight average molecular weights ranging from 10^3 to 10^5, and four samples of polymeric diallyldimethylammonium chloride (dadmac) with weight average molecular weights ranging from 10^4 to 10^5. These polymers are typically used for water clarification in waste treatment applications.

A correlation exists between the molecular weight and activity of polymers in wastewater treatment applications, but there is insufficient data to determine the effect of the molecular weight distribution on performance. Although the techniques used to determine the molecular weight distribution of polymers have been established for some time, the majority of the research has been conducted on non-aqueous polymers. Since most of the polymers used for wastewater treatment are water soluble, and often contain anionic or cationic functionalities, this research was an attempt to improve the database of molecular weight distribution information available for this important class of polymers.

The cationic polymers, epichlorohydrin/dimethylamine (epi/dma) and diallyldimethylammonium chloride (dadmac), are well known for their ability to clarify wastewater *(1)*. Previous work with these polymers has been performed with gel permeation chromatography (GPC) using calibration standards chemically different from the polymers being characterized (Bauer, D.S., Betz Chromatography Services, unpublished data.). In the present study, GPC was performed in conjunction with low angle laser light scattering (LALLS) to obtain absolute molecular weight distributions.

Experimental

Differential refractive indices (dn/dc) were measured with the LDC/Milton Roy Chromatix KMX-16 differential refractometer in the same solvent system as that used for the LALLS experiments. Four different concentrations of polymer were used and the differential refractive index values were averaged. If the dn/dc values changed with concentration, the y-intercept of a plot of dn/dc vs concentration was used in the molecular weight calculations.

Preliminary to running the GPC/LALLS experiments, the weight average molecular weights of the polymers were measured by using static LALLS in 1 M NaCl. These experiments were conducted using the Chromatix KMX-6 instrument from LDC/Milton Roy. The system parameters are contained in Table I. Two of the polymer samples were also measured in the 0.2 M LiNO3/0.1% TFA solvent system used for the GPC/LALLS experiment.

The GPC/LALLS instrumentation configuration consisted of an LDC Analytical ConstaMetric 3200 Bio Solvent Delivery System, Rheodyne Model 7125 Syringe Loading Sample Injector, LDC/Milton Roy Chromatix KMX-6 Low Angle Light Scattering Photometer, and LDC Analytical RefractoMonitor IV Refractive Index Detector. Data were collected and processed with the use of the LDC Analytical PCLALLS software package. Synchropak CATSEC columns were used in the following series: CATSEC 300 A (50 mm x 7.8 mm) guard column, CATSEC 4000 A (300 mm x 7.8 mm) column, CATSEC 300 A (30 mm x 7.8 mm) column.

The injector sample loop volume was calibrated by injecting a known concentration of sodium benzoate into a volumetric flask, filling to the mark, and measuring the UV-VIS absorbance. The absorbance was compared to a calibration curve to obtain the volume of injected sodium benzoate. The interdetector volume was calibrated by measuring the time difference between the LALLS and RI detector responses for an unfractionated polymer. The pump was calibrated daily by measuring the time required to fill a volumetric flask.

In order to avoid complications from impurities in the sample, the dadmac samples were dialyzed to remove any residual synthesis process salts. The solvent systems were prepared using doubly distilled water. For the GPC/LALLS experiments, a mobile phase of 0.1% trifluoroacetic acid (TFA) and 0.2 M LiNO3 was chosen (Bauer, D.S., Betz Chromatography Services, unpublished data.). D. J. Nagy et al. have shown universal calibration behavior with cationic polymers in a similar solvent system (0.1% TFA/0.20 N NaNO3) using CATSEC columns (2).

The intrinsic viscosities were measured by the Chromatography Services Group at Betz using the Schott-Gerate Automatic Viscometer.

Results

Epi/dma Polymers. Table II summarizes the numerical results for the epi/dma polymers. From Table II, it is evident that the Mw values obtained from the GPC/LALLS experiments with 0.2 M LiNO3/0.1% TFA, are consistently lower than

Table I. LALLS System Parameters

	Static LALLS	GPC/LALLS
Attenuator Constant	1.7703e-8	1.7703e-8
Fieldstop	0.15	0.15
Annulus	6-7 degrees	6-7 degrees
Cell Length	15 mm	4.93 mm
Solvent	1.0 M NaCl	0.2 M LiNO3/0.1% TFA
Refractive Index	1.343	1.335
Scattering Angle	4.835	4.865
1/(sigma'l')	881.33	877.36
Interdetector Volume		57 microliters
Flow	~0.2 mL/min	~0.6 mL/min
Loop Volume		60.5 microliters

Table II. Epi/dma Results

Polymer Sample	Intrinsic Viscosity	Static LALLS Mw	GPC LALLS Mw	Mz	Mn
A	0.22	324,000(+/-4,000)	235,000(+/-20,000)	1,810,000(+/-280,000)	15,300(+/-980)
B	0.16	43,800(+/-500)	36,100(+/-3,200)	131,000(+/-17,000)	10,800(+/-950)
C	0.13	30,100(+/-400)	19,700(+/-2,200)	41,100(+/-8,200)	6,900(+/-2,600)
D	0.05	8,800(+/-600)	4,180(+/-420)	15,000(+/-2,200)	1,880(+/-220)
E	0.03	3,510(+/-20)	2,350(+/-590)	3,280(+/-650)	1,450(+/-520)

the values obtained using static LALLS with 1 M NaCl. Since the hydrodynamic volume of a polymer may be influenced by the ionic strength of the solvent (3,4), it is not surprising that the values obtained from these two experiments should differ. Subsequent static LALLS measurements using 0.2 M LiNO$_3$/0.1% TFA indicate that lower weight average molecular weight values are obtained with this solvent system compared to the NaCl solvent system (Table III). With the same solvent system, the GPC/LALLS and static LALLS values are within experimental error.

Figures 1-5 show the unusual curve shapes seen for some of these polymers. The LALLS curve of Polymer A seems to indicate that the polymer is being excluded from the column pores, but the sharp rise in the LALLS peak is not duplicated in the RI peak. Polymers D and E appear to be mixtures of different low molecular weight batches. Polymer D has three distinct peaks in the LALLS curve and a shoulder on the RI curve. The LALLS data of Polymer E show one large peak and another small peak with a much higher molecular weight. The RI data does not show this extra peak, possibly because the concentration is too low. Both Polymers D and E are difficult to measure using this method because the LALLS technique is not very sensitive to low molecular weight polymers. Polymers B and C have nearly gaussian curve shapes.

Dadmac Polymers. Table IV summarizes the results for the dadmac polymers. Figures 6-9 show typical chromatograms for these samples. As with the epi/dma polymers, the weight average molecular weight values from the LiNO$_3$ solvent system are lower than the values obtained using NaCl (Table III). Except for the very low molecular weight sample, E, the reproducibility of the epi/dma polymers was fairly good (+/- 10%) from run to run, and from day to day. The dadmac polymer data showed greater variation and concentration dependence. Mz and Mw increased with increasing concentration, while Mn decreased. The solvent system for this particular polymer may not be optimized. Future work will include the use of different solvent systems to obtain more consistent, reproducible results.

Intrinsic Viscosities. Figure 10 shows the correlation between intrinsic viscosity and weight average molecular weight. A greater increase in intrinsic viscosity is seen with increasing molecular weight for the dadmac polymers, compared to the epi/dma polymers. This Figure also shows the trend toward higher molecular weight values measured in 1 M NaCl (solid symbols) compared to 0.2 M LiNO$_3$/0.1% TFA (hollow symbols). For both the epi/dma and dadmac polymers, the log of molecular weight is nearly linear with the log of intrinsic viscosity, but the data show some curvature. This curvature may be due to changes in the polymer conformation with increasing molecular weight. The linear correlation coefficients and the Mark Houwink constants are summarized in Table V. The Mark Houwink constants for the two solvent systems are within experimental error of each other and of the theoretical range predicted by Flory (5).

Table III. Comparison of Different Solvent Systems

Polymer Sample	1 M NaCl Static LALLS	0.2 M LiNO$_3$/0.1% TFA Static LALLS	0.2 M LiNO$_3$/0.1% TFA GPC/LALLS
C	30,100 (+/-400)	20,400 (+/-400)	19,700 (+/-2,200)
X	337,000 (+/-14,000)	213,000 (+/-18,000)	234,000 (+/-19,000)

Table IV. Dadmac Results

Polymer Sample	Intrinsic Viscosity	Static LALLS Mw	GPC LALLS Mw	Mz	Mn
W	0.99	703,000(+/-7,000)	418,000(+/-148,000)	952,000(+/-476,000)	89,000(+/-14,000)
X	0.90	337,000(+/-14,000)	234,000(+/-19,000)	716,000(+/-66,000)	42,700(+/-2,800)
Y	0.52	191,000(+/-1,000)	73,500(+/-9,000)	132,000(+/-18,000)	30,200(+/-5,400)
Z	0.15	15,700(+/-800)	11,900(+/-3,400)	19,800(+/-9,000)	7,000(+/-1,900)

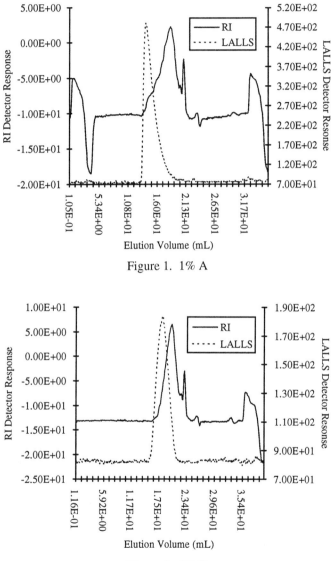

Figure 1. 1% A

Figure 2. 1% B

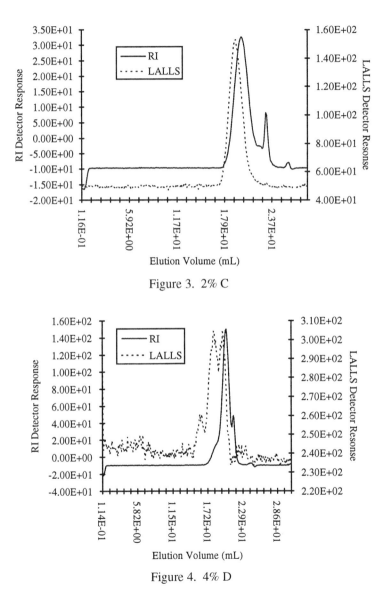

Figure 3. 2% C

Figure 4. 4% D

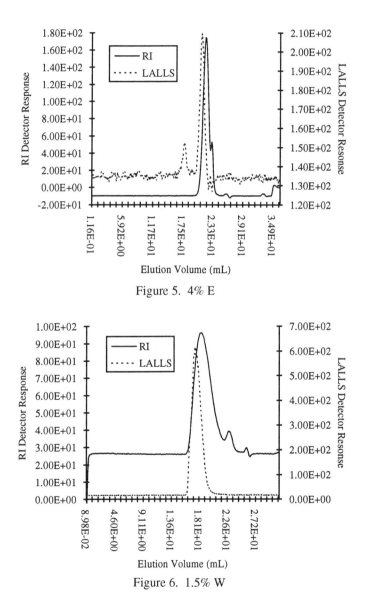

Figure 5. 4% E

Figure 6. 1.5% W

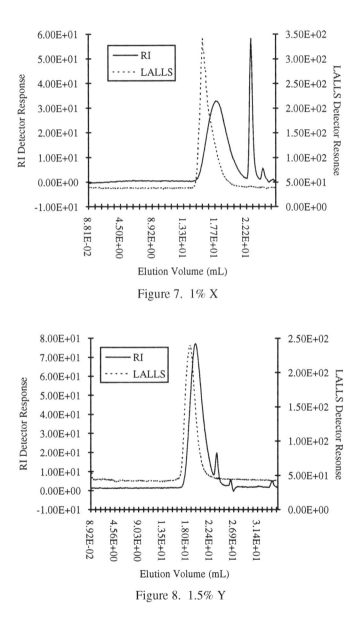

Figure 7. 1% X

Figure 8. 1.5% Y

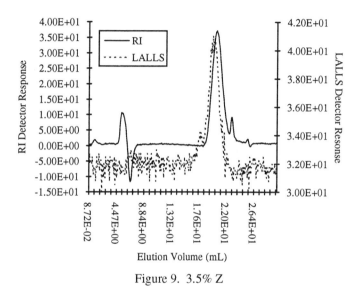

Figure 9. 3.5% Z

Table V. Least Squares Analysis of Log [η] vs Log Mw

Polymer	Static LALLS 1 M NaCl EPI/DMA	GPC/LALLS 0.2 M LiNO$_3$ EPI/DMA	Static LALLS 1 M NaCl DADMAC	GPC/LALLS 0.2 M LiNO$_3$ DADMAC
a	0.459 (+/-0.093)	0.434 (+/-0.082)	0.521 (+/-0.055)	0.548 (+/-0.063)
Log (K)	-3.06 (+/-0.42)	-2.87 (+/-0.35)	-3.01(+/-0.29)	-3.02 (+/-0.32)
corr.coeff.	0.94	0.95	0.99	0.99

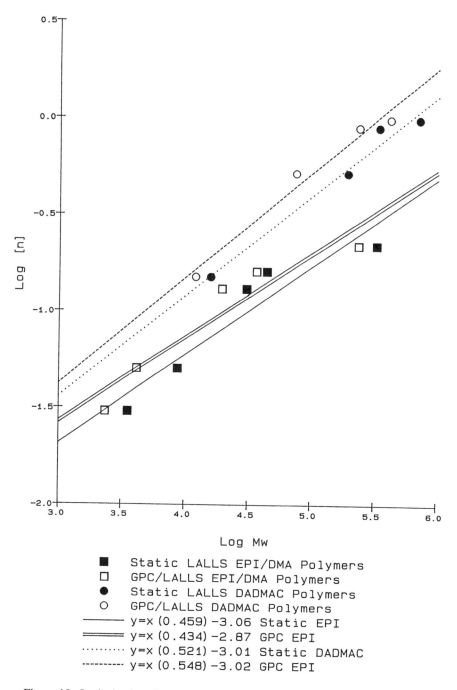

Figure 10. Intrinsic viscosity as a function of molecular weight, EPI/DMA and DADMAC polymers

Discussion

For static LALLS, 1.0 M NaCl was chosen as the solvent for two reasons: consistency with historical data generated on intrinsic viscosity measurements, and to minimize polymer extension. It is well established that in the presence of sufficient electrolyte polyelectrolytes will behave as neutral macromolecules. The solvent system chosen for GPC/LALLS was based on two criteria as well: the elimination of ion exclusion effects and the minimization of column corrosive effects caused by the presence of chloride ions (6).

The static LALLS Mw values measured in 1 M NaCl are higher than those generated by static LALLS in 0.2 M LiNO3/TFA or GPC/LALLS in 0.2 M LiNO3/TFA (the latter two are in agreement), but comparison of the 'a' constants for the Mark Houwink equation do not indicate any coil expansion as might be expected. A possible explanation may involve comparison of both the concentration and the nature of the solvent system.

When a salt is added to a polyelectrolyte solution, the counterions in the coils screen the cationic sites and the electrostatic interactions cause a reduction of the hydrodynamic dimension of the polymer. Based on these effects, the more highly concentrated 1 M NaCl solution is expected to exhibit a lower Mw than that measured in 0.2 M LiNO3. It has also been shown by Boussouira et al. (7) that for cationic ammonium copolymers, increased shielding of the polyion occurs when NO3$^-$ is used relative to Cl$^-$. This behavior was in agreement with the affinity order for an ammonium type anion exchanger. The increased shielding would result in a more tightly coiled polymer, which would manifest itself as lower Mw. Coupled with the effects the TFA and lower pH, the increased shielding ability of NO3$^-$ may account for the lower Mw measured in this solvent system. The competing effects of the concentration and the nature of the solvent system may then explain why the Mark Houwink constants for these two solvent systems are nearly identical. The influence of the solvent choice on the Mw will be explored in a future publication.

Conclusions

The technique of GPC/LALLS can be used to measure the molecular weight distribution of water soluble cationic polyelectrolytes typically used for industrial water treatment processes. Weight average molecular weights calculated from the GPC elution curves for epi/dma and dadmac polymers in 0.2 M LiNO3/0.1% TFA are consistently lower than those measured using 1 M NaCl. The different salts used in these two experiments are postulated to contribute to this effect.

Acknowledgments

The authors would like to acknowledge Dean S. Bauer of the Betz Chromatography Services Group for developing the buffer system used in these experiments, and thank him for his invaluable assistance throughout this project.

Literature Cited

(1) *Betz Handbook of Industrial Water Conditioning*; Moore, B. C., et al., Eds.; Ninth Edition; Betz Laboratories, Inc.: Trevose, PA, 1992; pp 24-25.

(2) Nagy, D. J.; Terwillinger, D. A.; Lawrey, B. D.; Tiedge, W. F. *In International GPC Symposium 1989 Proceedings*; Waters/Millipore Corporation: Milford, MA, 1989; pp 637-662.

(3) Potschka, M. J. *Chromatography* **1988**, *441*, pp 239-260.

(4) Herold, M. *American Laboratory* **1993**, *March*, pp 35-38.

(5) Flory, P. J. *Principles of Polymer Chemistry*; Cornell University Press: Ithaca, NY, 1953; pp. 280, 606.

(6) Gooding, D. L.; Schmuck M. N.; Gooding, K. M. *J. Liq. Chromatogr.* **1982**, *5*, 2259.

(7) Boussouira, B.; Ricard, A.; Audebert, R. *J. Poly. Sci.: Part B* **1988**, 26, 649-661.

RECEIVED August 6, 1993

Chapter 11

Chromatographic Characterization of Acrylic Polyampholytes

C. S. Patrickios, S. D. Gadam, S. M. Cramer, W. R. Hertler, and T. A. Hatton

Department of Chemical Engineering, Massachusetts Institute of Technology, Cambridge, MA 02139

The adsorption properties of synthetic polyampholytes based on methacrylic acid were investigated by gradient and frontal techniques on a strong anion-exchange column. These polyampholytes are of well-defined size (4,000 Da) and composition, contain low levels of impurities, and include random, diblock and triblock copolymers. While the random copolymers show no retention in the gradient elution and low adsorption in the frontal experiments, the block copolymers are strongly retained and adsorbed at large amounts. The diblock polyampholyte, which does not form micelles, exhibits similar adsorption properties as the triblock polyampholytes which form micelles. This suggests that the micellar character of the triblocks does not inhibit or reduce adsorption; on the contrary, the higher (than the single chain) micellar charge may provide a greater driving force for polymer adsorption.

Ion-exchange displacement chromatography of proteins is a separation technique in which a mixture of proteins is adsorbed on the column and subsequently displaced by a polyelectrolyte of higher column affinity, the displacer. This technique results in concentrated protein fractions and is therefore particularly suitable for separations of mixtures in which the desired proteins occur in very low concentrations. As displacement chromatography is gaining popularity, the quest for more efficient displacers is becoming necessary (1). We recently synthesized low-molecular-weight block polyampholytes based on methacrylic acid (Patrickios et al., MIT, submitted for publication). Initial experiments showed that these polymers can successfully displace and separate protein mixtures in ion-exchange columns. Most importantly these polymers possess two novel features: the column regeneration can be accomplished either by pH change or salinity increase, and any polymer contaminating the last protein fraction can be precipitated at the isoelectric point of the polymer.

These polyampholytes contain up to four different methacrylic residues presented in Figure 1: methacrylic acid (Ac) which can be negatively charged and has a pK of 5.4 (2), dimethylaminoethyl methacrylate (B) which can be positively charged and has a pK of 8.0 (2), methyl methacrylate (M) which is neutral and hydrophobic, and phenylethyl methacrylate (P) which is neutral and more hydrophobic than methyl methacrylate. The Group Transfer Polymerization (GTP) technique used for the

0097–6156/94/0548–0144$06.00/0

synthesis resulted in polymers with polydispersities as low as 1.3 and high composition homogeneity. The molecular weight of the polyampholytes is approximately 4,000 Da, with the exception of one polymer which is 15,000 Da. One neutral and one base-rich random polyampholytes were also synthesized. One polyampholyte is a neutral diblock (MW = 2,500 Da) and the rest are ABC triblocks with different acid/base ratio, hydrophobicity, and block sequence. Table I, in the Results and Discussion section, gives the composition and sequence of most of our copolymers. Both the diblock and the triblock copolymers precipitate around the isoelectric point. Light scattering studies revealed that at intermediate pH (=3-10) the triblocks, not the diblock, form micelles with hydrodynamic size larger than 10 nm. A steady-state pyrene fluorescence study on Polymer 6 (Table I) at pH 4.5 indicated a very low critical micellar concentration (below 0.1mg/mL) suggesting that, at the polymer concentrations employed in this study, typically 10 mg/mL, the triblock polyampholytes occur as micellar aggregates rather than free chains.

The aim of this study is to investigate the chromatographic behavior of these novel polyampholytic displacers in the absence of proteins. Most of the experiments were performed at pH 8.5 at which all of the polymers are soluble. The polymer parameters determined are the adsorptive capacity, the characteristic charge, and the steric factor (*3-6*). The characteristic charge is determined as the number of ionic bonds that a polymer forms with the stationary phase. The steric factor is the number of inaccessible column sites per polymer molecule at maximum (lowest salt concentration) column saturation.

Experimental

An analytical Waters Ion Exchange column of internal diameter 5 mm was packed to a length of 39 mm with 8 μm strong anion exchange (quaternary methylamine) beads of 100 nm average pore size. The same column was used for both the gradient elution and the frontal experiments. The equilibration buffer was Tris, typically at pH 8.5 and containing 50 mM Cl^-. A Waters Maxima 820 workstation was used for data acquisition.

Gradient Elution. A linear 10 min-gradient from 0.2 to 1.0M NaCl was applied at a flow rate of 0.5 mL/min using a Waters 600 Multisolvent Delivery System. The gradient delay was 9 min (due to the volume of the mixing chamber) and the dead volume of the column was 0.6 mL. A Waters 481 Lamda-Max LC Spectrophotometer was used to monitor the column effluent at 240 nm. It was not convenient to employ the wavelength of 310 nm used in the frontal experiments because the signal-to-noise ratio was very low. 20 μL of 10 mg/mL polymer samples prepared in Tris of pH 8.5 and 50 mM Cl^- were injected using a Rheodyne manual injector.

Frontal Experiments. Five frontal experiments (five steps) were performed for the characterization of each polymer at column saturation (*5*). An LKB 2150 HPLC pump was used for solvent delivery and a Spectroflow 757 detector was used to measure the absorbance of the effluent at 310 nm. A ten-port Valco manual injector with a 10 mL loop was used to inject the polymer, the nitrate, and the regenerant solutions. In the first frontal experiment (step 1), the column capacity in small anions was calculated by passing a front of 100 mM sodium nitrate at 0.5 mL/min through the equilibrated column and determining the nitrate breakthrough time. Second, after reequilibrating the column with buffer, a front of 10 mg/mL (2.5 mM) polyampholyte solution was passed at 0.2 mL/min, the polymer breakthrough time was determined and the amount of adsorbed polymer was calculated (step 2). The lower flow rate in step 2 secures low levels of pressure drop and adequate time for polymer adsorption. The non-adsorbed polymer in the dead volume was washed with buffer for 10 column volumes. Third, with the polymer adsorbed, a 30 mM sodium nitrate front was

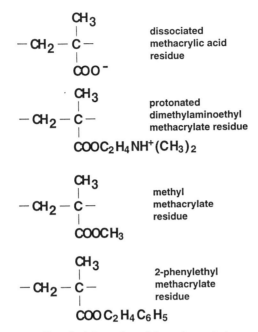

Figure 1. Chemical formulas of the polyampholyte residues.

Table I. Salt concentration required for polymer elution at pH=8.5

Polymer	Sequence	Composition	[NaCl][1] (mM)	# negative charges
1[2]	B-M-Ac	16-12-8	<200	8
2[2]	B-M-Ac	12-12-12	<200	12
3	M/Ac/B	12/12/12	247	12
4	B/Ac	10/10	298	10
5[3]	M/Ac/B	36/36/36	376	36
6	B/M/Ac	12/12/12	432	12
7	B/M/Ac	8/12/16	470	16
8	B/M/P/M/Ac	12/6/1/6/12	490	12
9	P/B/M/Ac	1/12/12/12	498	12
10	B/M/Ac	16/12/8	500	8
11	B/M/Ac	10/20/10	525	10
12	B/P/Ac	10/10/10	550	10

[1]salt concentration at peak maximum.
[2]random copolymers
[3]MW = 15,000Da

introduced at 0.2 mL/min, the nitrate breakthrough time was determined and the number of column sites not occupied by the polymer was calculated (step 3). The low sodium nitrate concentration was chosen so that the nitrate not displace any polymer. In the case of the one experiment at which the buffer used was only 5 mM Cl⁻, a 5 mM sodium nitrate concentration was used. Fourth, a 1M NaCl in 100 mM phosphate solution at pH 7.5 or 3.0 was introduced at 0.2 mL/min to desorb the displacer and regenerate the column (step 4). Fifth, after regeneration, the column was equilibrated with the buffer and a 100 mM sodium nitrate front was passed at 0.5 mL/min to test the regeneration efficiency by determining the nitrate breakthrough time and calculating the column capacity in nitrate (step 5).

In steps 2 and 3, the effluent between the column dead volume and the breakthrough volume was collected and analyzed for polyampholyte by gradient elution, and for chloride ions according to the ASTM assay (7). For calibration, 1 mL chloride standards in Tris buffers of different pH as well as in deionized water were transferred to 50 mL deionized water and titrated against 0.01 M silver nitrate using potassium chromate indicator solution. It was found that while standards in 50 mM Cl⁻ Tris buffers at pH 7.2 and 7.5 gave the same slope (μmoles of Cl⁻/mL of titrant) as standards in deionized water, standards in 50 mM Cl⁻ Tris buffers at pH 8.5 gave approximately twice the slope probably due to the increased concentration of Tris Amine at this pH which competes with silver ions for chloride. We also observed that the presence of the polycationic impurities in the polyampholytes caused sometimes different color changes in the assay.

Results and Discussion

Both the gradient elution and the frontal experiments were performed at concentrations of 10 mg/mL that belong to the non-linear part of the polymer isotherm (with the exception perhaps of the random copolymers). The non-linearity at 10 mg/mL was established in preliminary frontal experiments with triblock copolymers and showed that an increase in the feed polymer concentration from 10 to 50 mg/mL had no effect on the amount of polymer adsorbed. Polyelectrolytes of high charge density, such as DEAE-dextran or dextran sulphate, typically exhibit square isotherms with the linear part lying at concentrations below 1 mg/mL (5).

Gradient Elution. Figure 2 is the gradient elution chromatogram of the acid-rich triblock polyampholyte (Polymer 7 in Table I). Similar chromatograms were obtained for the other polymers listed in Table I. Polymer 7 comes out of the column after 14 min which corresponds to a salt concentration of 470 mM. The sharpness of this peak suggests that the polymer is homogeneous in composition. Besides the major peak, three small unretained peaks appear that add up to less than 5% of the area of the major peak. This means that the polymer is very pure. The peak at 1 min corresponds to the dead volume of the column and is probably a polycationic impurity (terminated first block and diblock). The other two peaks are probably polymer with a small number of negative charges (early terminated triblock).

The estimation of 95% purity of Polymer 7 is based on the assumption that the impurities and the pure polymer have similar extinction coefficients. If the extinction coefficient of the impurities is much smaller than that of the polymer, the purity is much lower than 95%. For this reason it was necessary to estimate the purity by a second method. Polymer was dissolved in acid solution and precipitated at the isoelectric point by addition of the appropriate volume of potassium hydroxide solution. This procedure results in the purification of the polymer because the impurities do not precipitate. The dissolution-precipitation cycle was repeated five times and, finally, the polymer was dried. A solution of the purified polymer was subjected to gradient elution analysis and the obtained chromatogram was compared to that of the unpurified polymer. The major peak of the unpurified polymer was 15%

smaller than that of the purified polymer both in terms of area and height. This suggests that the original sample contains 15% impurities which can still be considered a small contamination. These results are in agreement with the findings of Möller and coworkers who determined impuriry levels around 10% for their copolymers synthesized by GTP.

Table I summarizes the results of the gradient elution experiments. The retention time at the peak maximum was determined and converted to the corresponding salt concentration. The number of negative charges per polymer molecule (taken from the experimental titration curves) at the pH of the experiment also appears in Table I. By examining the Table, we can make four important observations.

First, the random copolymers (Polymers 1 and 2) are not retained, inspite of the fact that they have the same composition with block copolymers that are retained (Polymers 10, 3 and 6). This can be attributed to the random distribution of the adsorbing residues of the random copolymers which results in a lower local charge density. By comparing, for example, Polymers 2 and 3, we can estimate that the linear density in negative charges of the former is the one third of that of the latter.

Second, although the diblock copolymer (Polymer 4) and the random copolymers do not form micelles (Patrickios et al., MIT, submitted for publication), the diblock is retained. More specifically, it can be observed that it is retained more strongly than triblock Polymer 3 that does form micelles.

Third, although all the block polyampholytes are retained, there is no correlation between the retention and the length of the negative block. There is also no correlation between the retention and the net charge (not shown in the Table). There is, however, a strong effect of the hydrophobic block on retention. Focusing on Polymers 4, 6, and 11, which have similar lengths of negative and positive blocks but different lengths of the methyl methacrylate block in the middle, we can see that retention increases with the length of the methyl methacrylate block. This can be attributed to two effects: first, the middle block spaces away from the adsorbing surface the repelling amine block and, second, lateral middle block interpolymer hydrophobic interactions enhance retention. The comparison of the retentions of Polymers 11 and 12 points towards the greater importance of the hydrophobic interactions as the retention of Polymer 12, which bears the very hydrophobic phenylethyl methacrylate residues, is stronger, inspite of the shorter spacer length. Since the hydrophobic interactions are of short range and since the hydrophobic blocks of the adsorbed molecules are probably not in direct contact, one would dispute the two-dimensional hydrophobic interaction scheme. A different explanation can be given by considering the hydrophobic interactions in three dimensions, i.e. the micellization of the triblocks in solution. The polymer migrates down the column probably not as single chains but as micelles with a charge which is several times that of the chain monomer. It is likely that a more hydrophobic block will result in micelles with larger aggregation numbers and, therefore, higher micellar charge. It might be expected that retention would correlate well with the micellar charge.

Fourth, by examining Polymers 3 and 5, the significance of the position of the adsorbing block can be realized. These polymers show decreased retention because the adsorbing block lies in the middle of the molecule which leads to steric hindrance and decreased flexibility of the adsorbing block. Another negative factor is the close proximity of the repelling block to the adsorbing surface. Polymer 5 is retained more than Polymer 3 because it is three times larger.

The above interpretations on the order of elution are qualitative and are not based on any model. A more complete analysis is in progress and is based on isocratic elution of polymer samples at different salt concentrations which will result in the determination of the characteristic charge and the equilibrium constant (5). These two quantities define the affinity of the solute for the stationary phase according to the model of Brooks and Cramer (6). It is therefore likely that the block copolymers with similar adsorbing blocks (and probably similar characteristic charge) exhibited

different elution times because of different equilibrium constants. The equilibrium constant is expected to incorporate the effects of the length of the neutral block (hydrophobic interactions and spacing out of oppositely charged blocks) and of the block sequence, in addition to those of the lengths of the adsorbing and repelling blocks (electrostatic interactions). It is worth pointing out that the saturation capacities determined by the frontal analysis in the following section, performed mainly at a single set of conditions, are not expected to be influenced by the equilibrium constant but they should be dictated only by the characteristic charge and the steric factor (*6*). Consequently, it should not be surprising if hydrophobic interactions appear to play no role in these results.

Frontal Experiments. Figure 3 shows two typical polymer fronts at pH 8.5 and 50 mM Cl⁻ as monitored at 310 nm. The midpoint of the polymer breakthrough is very clear and can be used to calculate the amount of polymer adsorbed. The shallow breakthrough from 2.5 min (dead volume of the column) to the polymer breakthrough is due to displaced salt and unretained impurities. Samples collected from this volume and analyzed by gradient elution showed the absence polymer and the presence of unretained impurities. Also chloride analysis demonstrated the presence of chlorides at concentrations higher than that of the buffer. The exact amount of chlorides displaced by the polymer was determined from the chloride analysis. Assuming a stoichiometric model for polymer adsorption, this amount of chlorides corresponds to the number of bonds between the polymer and the column.

As already mentioned, after polymer adsorption and washing with buffer, the number of sites inaccessible to the polymer are calculated by passing sodium nitrate and determining the breakthrough volume of the front. Sodium nitrate was chosen as the nitrate ion has a high absorbance. Samples collected from this step and analyzed by gradient elution showed absence of polymer. This analysis confirmed that the nitrate does not displace any polymer, which was expected because the nitrate concentration was lower than that of the chloride in the buffer. As the nitrate front goes through the column, it displaces the chloride ions which are bound to the column sites which are inaccessible to the polymer. Samples collected from this step and analyzed for chloride resulted in the determination of a number of inaccessible column sites very similar to that determined by the nitrate breakthrough volume. The number of sites occupied by the polymer can be calculated by subtracting the number of inaccessible sites from the total small-ion column capacity which was approximately 132 μmoles.

One can argue that the nitrate front will also displace the chloride counterions of the positively charged residues of the adsorbed polyampholyte as well. This will result in an overestimation of the number of inaccessible sites and an underestimation of the number of occupied sites. We checked the extent of this error by comparing the number of occupied sites calculated in step 3 from the nitrate front with the number of occupied sites determined in step 2 from the chloride analysis. The difference was less than 5% which is within experimental error. This should have been expected because at the pH of most of the experiments (= 8.5), 40% of the amine residues are uncharged.

Regeneration in step 4 was always successful and it took place in less than two column volumes. At the beginning of regeneration the pressure drop rose up to 20 atm for 2-3 min due to the high concentration of the polymer that was released. The nitrate frontal experiment in step 5 showed that the ion capacity of the column was fully recovered.

Effect of Polymer Type. Table II summarizes the results obtained for different polymers at pH 8.5 and 50 mM Cl⁻ and includes the adsorptive capacity, the number of occupied sites as determined from the nitrate front of step 2, and the characteristic charge calculated as the number of occupied sites per adsorbed polymer molecule.

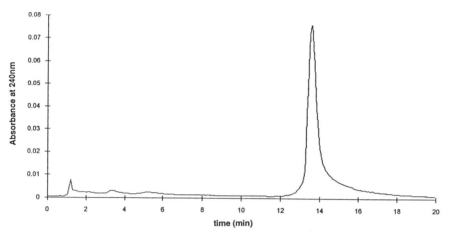

Figure 2. Elution of the acid-rich triblock polyampholyte by a linear 0.2-1.0M NaCl gradient at pH 8.5 and flow rate of 0.5 mL/min.

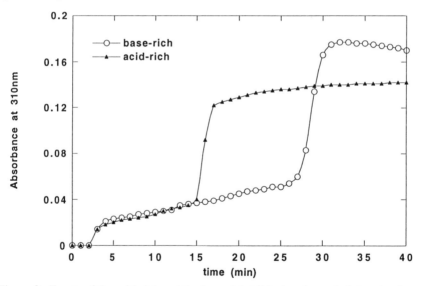

Figure 3. Fronts of the acid-rich and the base-rich triblock polyampholytes at polymer concentration 10 mg/mL, pH 8.5 and flow rate 0.2 mL/min.

Table II. Adsorptive capacity and characteristic charge for synthetic polyampholytes at pH = 8.5

Pol	MW	ads. (mg)	ads. (μmol)	unoc. sites (μmol)	occ. sites (μmol)	char. charge	# neg. charges	fraction of groups bound
7	3832	24.5	6.39	78.2	53.8	8.4	16	0.53
6	4116	31.0	7.52	78.9	53.1	7.1	12	0.59
4	2430	16.1	6.63	81.9	50.1	7.6	10	0.76
2	4116	12.0	2.91	104.9	27.1	9.3	12	0.78
10	4400	49.9	11.3	82.1	49.9	4.4	8	0.55

By examining the fourth column of the Table it can be observed that, with the exception of random Polymer 2, the number of polymer molecules adsorbed increases as the number of negative charges per molecule decreases. By examining now the sixth column we can see that, again with the exception of the random polymer, the number of occupied sites is always the same, independent of polymer composition. This can be understood because all the polymers have the same adsorbing block, poly(methacrylic acid), located at the end of the molecule. The number of occupied column sites does not appear to be affected by the different lengths of the positive block or the absence of hydrophobic block in Polymer 4. A recent frontal experiment with poly(methacrylic acid) (an oligomer of 12 units) showed that the homopolymer occupied the same number of column sites as the block copolymers. This confirmed that, at this pH, the non-adsorbing blocks do not affect the adsorption and probably extent vertically to the adsorbing surface as shown in Figure 4. The calculated characteristic charge of the block copolymers follows the same trend as the number of negative charges obtained from the experimental titration curves. The random copolymer has the negative charges randomly distributed and "mixed" with the positive ones. This results in a weak driving force for adsorption and in a flat adsorption conformation. These lead finally to the small amount of polymer adsorbed and the smaller number of column sites occupied per molecule.

Effect of pH. Table III shows the results for the adsorption of the acid-rich polymer at pH from 7.2 to 8.5. It was not possible to go to lower pH because the polymer precipitates. We can see that more polymer molecules are adsorbed as the pH is lowered. For the pH range studied the number of occupied sites does not appreciably change and the experimental hydrogen-ion titration curve suggests that the number of negative charges of the polymer is almost constant. On the other hand, by decreasing the pH from 8.5 to 7.2, the hydrogen-ion titration curve suggests that the number of positive charges per molecule increases from 5 to 8. It seems, therefore, that the stronger repulsion between the matrix and the amine block at lower pH allows only to a smaller number of negative residues per molecule to interact with the adsorbing surface. Since the number of occupied sites is constant, more polymer molecules per column area will adsorb.

Effect of salt concentration. We performed one experiment with Polymer 6 at the very low chloride concentration of 5 mM at pH 8.5 in order to test whether the 50 mM chloride concentration, used in all the previous experiments, was low enough to lead to maximum column occupancy. The results, listed in Table IV, indicate that, at the lower ionic strength, more polymer is adsorbed and more column sites are covered. This can be attributed to the weaker screening of the electrostatic interactions between the matrix and the polymer at the lower salt concentration. The constant value of the characteristic charge at the two chloride concentrations suggests that the adsorption conformation remains the same. Assuming now that the column occupancy will not increase further by going to even lower chloride concentrations, we calculate the steric factor for the polymer as the number of inaccessible sites per adsorbed polymer molecule at 5 mM chloride. The steric factor equals 7.3 and the ratio of the steric factor to the characteristic charge equals 7.3/7.2 = 1.01 which is very similar to that of dextran sulphate (5). A similar result was obtained in the frontal experiment with the oligo(methacrylic acid) giving again a steric factor to characteristic charge ratio close to one. This is in agreement with the very low (5%) isotacticity of polymethacrylates synthesized by GTP at room temperature (9) which implies that only 50% of the carboxylates can be oriented towards the adsorbing surface.

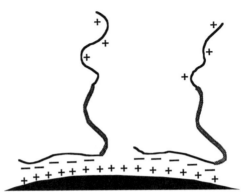

Figure 4. Adsorption conformation of a triblock polyampholyte at pH 8.5.

Table III. Effect of pH on the adsorptive capacity and characteristic charge of the acid-rich triblock polyampholyte (Pol 7)

pH	ads. (mg)	ads. (μmol)	unoc. sites (μmol)	occ. sites (μmol)	char. charge	# neg. charges	fraction of groups bound
7.2	41.5	10.8	75.0	57.0	5.3	15	0.35
7.5	38.5	10.0	68.1	63.9	6.4	16	0.40
8.5	24.5	6.4	78.2	53.8	8.4	16	0.53

Table IV. Effect of salt concentration on the adsorptive capacity and characteristic charge of the neutral triblock polyampholyte (Pol 6) at pH = 8.5

[Cl$^-$] (mM)	ads. (mg)	ads. (μmol)	unoc. sites (μmol)	occ. sites (μmol)	char. charge	steric factor
5	37.5	9.1	66.2	65.8	7.2	7.3
50	31.0	7.5	78.9	53.1	7.1	---

Conclusion

The chromatographic techniques of this investigation are powerful tools for providing an understanding of the complicated behavior of our block polyampholytes in two dimensions. At low loading, the polymer affinity to the column is enhanced by the hydrophobic interactions. At column saturation and pH 8.5, the adsorptive capacity is dictated by the size of the adsorbing methacrylic acid block and not by the amine block which is only partially charged (Figure 4). At column saturation and at lower pH, the ads.orptive capacity is also influenced by the amine (repelling) block which gets fully charged. Further studies will result in a more complete understanding of the adsorption behavior of these polymers and provide the ability to predict their protein displacing characteristics.

Literature Cited

1. Cramer, S. M. *Nature* **1991**, *351,* 251.
2. Merle Y. *J. Phys. Chem.* **1987, *91,* 3092.
3. Kopaciewicz, W.; Rounds, M. A.; Fausnaugh, J.; Regnier, F. E. *J. Chromatogr.* **1983,** *266,* 3.
4. Jen, S. C. D.; Pinto, N. G. *J. Chromatogr. Sci.* **1991,** *29,* 478.
5. Gadam, S. D.; Jayaraman, G.; Cramer S. M. *J. Chromatogr.* **1993,** *630,* 37.
6. Brooks, C. A.; Cramer, S. M. *AIChE J.* **1992,** *38,* 1969.
7. ASTM *Annual Book of ASTM Standards* Philadelphia, PA, 1991; 11.01.
8. Möller, M. A.; Augenstein, M.; Dumont, E.; Pennewiss, H. *New Polymeric Mater.* **1991,** 2, 315.
9. Sogah, D. Y.; Hertler, W. R.; Webster, O. W.; Cohen, G. M. *Macromolecules,* **1987,** *20,* 1473.

RECEIVED August 30, 1993

Chapter 12

Determination of Microdomain Size of Hydrophobic Polyelectrolytes by Luminescence Quenching

Ulrich P. Strauss, Vincent S. Zdanowicz, and Yuanzhen Zhong[1]

Department of Chemistry, Rutgers, The State University of New Jersey, New Brunswick, NJ 08903

The luminescence decay kinetics of a ruthenium complex solubilized by a hydrolyzed copolymer of maleic anhydride and n-hexyl vinyl ether (DP 1100) were measured at different pHs as a function of the concentration of quencher, 9-methyl anthracene, in order to determine the size of the intramolecular micelles formed by the copolymer. The solubilization of quencher was used to determine the degree of micellization of the copolymer at various pHs. A time-resolved, single photon counting technique was employed for the luminescence decay measurements. Results indicated that an average of 42 repeat units comprised a micelle, indicating that one copolymer chain can form many intramolecular micelles. The micelle size was independent of pH and quencher concentration.

The hydrolyzed alternating copolymers of maleic anhydride and alkyl vinyl ethers have been useful for exploring the effects of hydrophobicity on the behavior of polyelectrolytes (*1-4*). With these copolymers the hydrophobicity is determined by the size of the alkyl group, while the charge density may be controlled by varying the pH. When the alkyl group contains less than four carbon atoms, the copolymers behave as normal polyacids; when the alkyl group contains twelve or more carbon atoms, the copolymers behave as typical polysoaps, showing the characteristic hypercoiled compact structures brought about by hydrophobic interactions between the alkyl groups. Perhaps the most interesting behavior is exhibited by the copolymers with alkyl group sizes in the range between four and ten carbon atoms: here we observe typical polysoap behavior at low pH and normal polyelectrolyte behavior at high pH, connected by a conformational transition reminiscent of denaturation phenomena exhibited in the realm of biological macromolecules (*1*).

One of the questions that arose in the course of these investigations was

---

[1]Current address: International Specialty Products, 1361 Alps Road, Wayne, NJ 07470

whether the hypercoiled state was characterized by one large compact cluster or by several smaller intramolecular micelles. The rather small cooperative unit size deduced from the pH dependence of the conformational transition (*5*) suggested the latter. This conclusion was later confirmed by aggregation number determinations employing luminescence quenching (*2*). The basis for this technique is to deduce the number of micelles from the probability of a quenching event: if both probe and quencher are solubilized by the micelles, the probability of probe and quencher molecules being located simultaneously in the same micelle will decrease with increasing number of micelles (*6*). This method is especially useful for intramolecular micelles for which more conventional techniques, such as light scattering or hydrodynamic procedures requiring that the micelles occur as independent units, are not applicable (*2,4,7*).

The system investigated previously in this laboratory consisted of a hydrolyzed copolymer of maleic anhydride and n-hexyl vinyl ether (hexyl copolymer) with degree of polymerization 1700, tris(2,2'-bipyridyl)ruthenium(II) ion, $[Ru(bpy)_3^{2+}]$, as the probe, 9-methylanthracene (9-MeA) as the quencher, and an aqueous 0.1 M LiCl solution as the solvent (*2*). The microdomains were found to encompass approximately 24 repeat units, unaffected by the concentration or by the extent of micellization of the polyacid. Neither did varying the probe concentration have any effect on this number. However, the previous investigation suffered from the disadvantage that the light source of the pulse sampling apparatus used had a 14 ns pulse width, too large to allow a direct determination of the initial luminescence intensity, necessary for the complete determination of all the physically significant parameters. Consequently, the dynamic method had to be supplemented with data obtained with a steady state luminescence instrument. The lack of perfect matching of the light sources and some of the assumptions needed in combining the data from the two instruments may have resulted in errors of unknown magnitude.

When a more satisfactory, time-resolved, single photon counting instrument became available to us, we resumed the measurements on similar systems, involving a hexyl copolymer with the same probe, quencher and solvent used previously. With this instrument, the pulse width of the light source was narrow enough (< 2 ns) to allow acquisition of all necessary decay parameters. In this new investigation we also used an improved method for measuring the solubilization limit of the quencher by the polyacid. This quantity is needed for the determinations of both the fraction of polymer in micellar form and the distribution of quencher between the polymer molecules and the solvent medium. Previously the solubilization was obtained by monitoring the increase in the turbidity caused by increasing amounts of quencher beyond its solubility limit; in the new study, that method was replaced by measurements of the optical density originating from the solubilized quencher at its solubility limit. The results of this new investigation will be presented below.

Experimental

Materials. The copolymer of maleic anhydride and hexyl vinyl ether was our sample BRB #2. Its degree of polymerization was estimated from viscosity measurements to be 1100. Its concentration, C_p, is expressed in moles of repeat units per liter. 9-Methylanthracene was recrystallized from ethyl alcohol. It was brought to the desired concentrations in the polymer solutions from alcoholic stock solutions as described previously (2). Tris(2,2'-bipyridyl)ruthenium(II) chloride hexahydrate was used as received.

Methods. Luminescence lifetime measurements were conducted using a Photochemical Research Associates (PRA) Model 3000 Fluorescence Lifetime Instrument (FLI), configured for single photon counting. The configuration and operation were similar to those described by Snyder et al (8). Luminescence decay data were analyzed using software obtained from Photon Technologies International (PTI), which utilizes an iterative reconvolution procedure employing a weighted, non-linear, least squares curve fitting method. The decay function used was

$$I(t) = A_1 \exp\left[-A_2 t - A_3\left(1 - \exp\left(-A_4 t\right)\right)\right] \tag{1}$$

where I(t) is the luminescence intensity (9). The criteria used to judge the goodness of fit were the reduced Chi-squared value, the Durbin-Watson parameter, the Runs Test statistic and plots of weighted residuals and the autocorrelation function.

The solubility limit of 9-MeA in polymer solutions was determined from a plot of the absorbance at 333 nm against the concentration of 9-MeA. Up to the solubility limit, the data follow Beer's Law. Beyond the solubility limit, the points lie on a horizontal line, if sufficient time is allowed for the excess 9-MeA to precipitate out of solution. The intersection between the two lines gives the limit of solubility. The absorbance measurements were performed with a Hewlett Packard 8450 Diode Array Spectrophotometer.

Results and Discussion

Solubilization of Quencher. Values of the solubilization, S, expressed in moles of 9-MeA solubilized per mole of polymer repeat unit, are shown in Table I as a function of pH. The values of S were obtained from the measured solubilities in polymer solutions, corrected for the solubility in the 0.1 M LiCl solvent, which we determined to be 3.0 x 10^{-6} M. As had been observed previously, the values of S decrease with increasing pH, reflecting a corresponding decrease in the extent of micellization of the polyacid (10). Since we found that the solubilization does not increase further as the pH is decreased below 4.2, we assume that micellization has reached its maximum at this point and that Θ_m, defined as the fraction of repeat units in micellized form, is equal to unity at pH 4.2 and below. By taking Θ_m proportional to S, we obtain the values given in the last column of Table I.

It has been shown experimentally that the distribution of solubilized molecules, in this case the quencher molecules, among micelles is given by a mass action law of the form

$$K' = \frac{[Q_m]}{[Q_a]\,\Theta_m\,C_P} \tag{2}$$

Table 1. Solubilization of 9-Methylanthracene as Function of pH

pH	$S (\times 10^3)^a$	Θ_m
4.2	8.62	1.000
4.5	8.34	0.967
4.8	7.54	0.875
5.1	6.75	0.783
5.4	5.61	0.651
5.7	5.17	0.600

a. Moles of 9-MeA per mole of polyacid repeat unit.

where $[Q_m]$ and $[Q_a]$ are the molar concentrations of solubilized and free quencher, respectively (11). This law can be derived with the assumption that the presence of solubilized guest molecules in a micelle does not affect the entry rate of further guest molecules (4). Since the equation is valid below, as well as at, the solubility limit, we can make use of the fact that in the latter case $[Q_m] / \Theta_m C_p = S$ and $[Q_a]$ is the solubility of the quencher in the solvent. We then find that $K' = 2870 \, M^{-1}$. This value, together with the known total concentration of quencher, allows the calculation of $[Q_m]$ and $[Q_a]$ by means of equation 2.

Luminescence Quenching. The time dependence of the luminescence of a probe in a micellar environment containing mobile quenchers is given by equation 1 above. The coefficients of this equation are related to the molecular parameters of interest by the relations (9)

$$A_2 = k_0 + \frac{k_Q \, k_-}{k_Q + k_-} \frac{[Q_m]}{[M]} \tag{3}$$

$$A_3 = \left(\frac{k_Q}{k_Q + k_-}\right)^2 \frac{[Q_m]}{[M]} \tag{4}$$

$$A_4 = k_Q + k_- \tag{5}$$

where k_0 is the first-order rate constant for the luminescence decay of the probe in the absence of quencher, k_Q is the first-order rate constant for quenching inside a micelle, k_- is the exit rate constant for a quencher from a micelle, and $[M]$ is the concentration of micelles expressed in moles per liter.

The procedure used for obtaining the desired molecular parameters by means of equations 3 - 5 is illustrated with a sample run involving five solutions differing only in their quencher concentrations. The relevant data are given in Table II. First, the derivatives of A_2 and A_3 with respect to $[Q_m]$ are determined by linear, least squares fitting. If the ratio of these derivatives is denoted by D, k_- is obtained by the relation

$$k_- = \frac{D}{1 + (D/A_4)} \tag{6}$$

which, together with the average value of A_4, leads to k_Q via equation 5. Next, the micelle concentration, [M], is determined from $dA_3/d[Q_m]$ by means of equation 4. Finally, the micelle size, n, defined as the average number of hexyl groups per micelle, is obtained from the expression

$$n = \Theta_m \, C_P \, / \, [M] \tag{7}$$

Results obtained under different experimental conditions and calculated in the manner illustrated above are compared in Table III. We have also included values of the equilibrium constant, K, for the distribution between micelles and solvent, given by the relation

$$K = \frac{[Q_m]}{[Q_a] \, [M]} \tag{8}$$

and calculated more efficiently by the equation K = nK'.

The values of n are substantially larger and the values of k_Q and k_- substantially smaller than those found in our previous investigation. We ascribe these differences to the improvements in the experimental methods discussed above. The new values of k_Q are also smaller than those found with micelles of simple surfactants, such as sodium dodecyl sulfate, but are of the same magnitude as those observed by others for intramolecular micelles attached to macromolecules. This effect has been attributed to an enhancement of the intramicellar viscosity caused by the attachment of the hydrocarbon groups to the polymeric backbone (4, 12). It has been shown previously that the values of k_- cannot be solely attributed to the transfer of quencher from micelles into the aqueous phase (2). The argument previously made to reach this conclusion is also supported by the present data in that they show that the hypothetical value of the second order rate constant for a quencher entering a micelle from the aqueous phase, k_+, given by the product of K and k_-, is too large to be theoretically feasible. The new data, however, are quite compatible with the hypothesis of a direct exchange of quencher between micelles, for which several mechanisms have been proposed (13), and which should be facilitated by the close proximity of micelles belonging to the same macromolecule.

Table II. Parameters of Sample Run [a]

[Q]	A_2	A_3	A_4	$dA_2/d[Q_m]$	$dA_3/d[Q_m]$	k	[M]	n
$\times 10^4$	$\times 10^{-6}$ (s^{-1})		$\times 10^{-6}$ (s^{-1})	$\times 10^{-8}$ (M^{-1}s^{-1})	(M^{-1})	$\times 10^{-5}$ (s^{-1})	$\times 10^4$	
0.00	1.196							
0.69	1.230	0.068	4.66					
1.37	1.259	0.145	3.98					
2.05	1.271	0.229	3.16					
2.73	1.290	0.307	3.03					
			3.93[b]	2.85	1190	2.3	7.45	44

a. C_p = 0.0538 M; pH = 5.7; Θ_m = 0.600; [Ru(byp)$_3$Cl$_2$] = 1.08$\times 10^{-5}$ M; [Q$_m$] / [Q] = 0.989.
b. Average A_4.

It is worth noting that the calculation of n is based on the assumption that the probe is completely solubilized in the micelles. Any probe molecules located elsewhere would, for all practical purposes, not be subject to quenching by the 9-methylanthracene, which is known to be 99% solubilized in the micelles. The resulting decrease in quenching would produce an apparent increase in the calculated micelle concentration, [M], and, hence, an apparent decrease in the micelle size, n. However, it has been shown previously by fluorescence experiments that the probe, even though cationic, is solubilized almost exclusively in the micellar regions (2), thus validating the procedure used here. Further validation of the approach taken here was obtained by the use of a different probe-quencher combination with the same polymer-solvent system (14). With pyrene as the probe and nonyl phenyl ketone as the quencher, both of which were shown to be solubilized exclusively in the micelles, the value of n obtained was 45 ± 1, which is within the upper range of the data given in Table III.

Published accounts of polymer-based micelles are scarce, and difficult to compare with our results. Chu and Thomas reported that the micelle sizes of two maleic acid-octadecene copolymers, one of DP 24 (12) and one of DP 17 (15), were equal to their respective degrees of polymerization. In contrast to our copolymer, their whole polymer molecules were smaller than one of our micelles. Binana-Limbelé and Zana (4) studied a high molecular weight copolymer of maleic acid and

Table III. Micelle Size and Related Parameters

pH	C_P	$[Ru(byp)_3^{2+}]$	Θ_m	k_0	k_Q	k_-	K	n
	$\times 10^2$	$\times 10^6$		$\times 10^{-6}$ (s^{-1})	$\times 10^{-6}$ (s^{-1})	$\times 10^{-5}$ (s^{-1})	$\times 10^{-4}$	
4.5	5.21	9.83	0.967	1.21	3.77	3.4	12.3	43
4.8	4.90	9.02	0.874	1.11	1.67	5.4	13.5	47
5.1	4.79	9.02	0.782	1.13	1.76	3.9	10.9	38
5.7	4.53	9.02	0.600	1.17	4.34	3.1	10.3	36
5.7	5.38	10.8	0.600	1.20	3.70	2.3	12.6	44
Ave:				1.16	3.05	3.6	11.9	42

decyl vinyl ether and reported a decrease in the micelle size from 79 to 15 with increasing degree of ionization, in contrast to our results which showed no such change. However, their calculations contained the implicit assumption that their quencher, dodecyl pyridinium chloride, was bound to the polymer exclusively in its micellar regions. It is likely that the cationic dodecyl pyridinium ions are also bound to non-micellar anionic polymer groups. Knowledge of the actual quencher distribution would strongly affect the calculated micelle size and its dependence on the degree of ionization. Resolution of this question requires experimental determination before a quantitative comparison between our results and theirs can be made. Nevertheless, whatever the outcome of such a determination, their conclusion, in agreement with our findings, that a large polysoap molecule contains a great number of small micellar microdomains, will retain its validity.

References

1. Dubin, P.; Strauss, U. P. *J. Phys. Chem.* **1970,** *74,* 2842.
2. Hsu, J.-L.; Strauss, U. P. *J. Phys. Chem.* **1987,** *91,* 6238.
3. Ito, K.; Ono, H.; Yamashita, Y. *J. Colloid Sci.* **1964,** *19,* 28.
4. Binana-Limbele, W.; Zana, R. *Macromolecules.* **1990,** *23,* 2731.
5. Barbieri, B. W.; Strauss, U. P. *Macromolecules.* **1985,** *18,* 411.
6. Turro, N. J.; Yekta, A. *J. Am. Chem. Soc.* **1978,** *100,* 5951.
7. Olea, A. F.; Thomas, J. K. *Macromolecules.* **1989,** *22,* 1165.
8. Snyder, S. W.; Demas, J. N.; DeGraff, B. A. *Anal. Chem.* **1989,** *61,* 2704.

9. Tachiya, M. *Chem. Phys. Lett.* **1975**, *33*, 289.
10. Strauss, U. P.; Schlesinger, M.S. *J. Phys. Chem.* **1978**, *82*, 1627.
11. Almgren, M.; Grieser, F.; Thomas, J. K. *J. Am. Chem. Soc.* **1979**, *101*, 279.
12. Chu, D.-Y.; Thomas, J. K. *Macromolecules.* **1987**, *20*, 2133.
13. Malliaris, A.; Lang, J.; Zana, R. *J. Phys. Chem.* **1986**, *90*, 655.
14. Zdanowicz, V. S.; Strauss, U. P. *Macromolecules.* In press.
15. Chu, D.-Y.; Thomas, J. K. In *Polymers in Aqueous Media: Performance through Association*; Glass, J. E., Ed.; Advances in Chemistry Series 233; American Chemical Society: Washington, DC, **1989**, pp 325-341.

RECEIVED August 6, 1993

Chapter 13

Characterization of Polyacrylamide-*co*-Sodium Acrylate

D. Hunkeler[1], X. Y. Wu[2], A. E. Hamielec[2], R. H. Pelton[2], and D. R. Woods[2]

[1]Department of Chemical Engineering, Vanderbilt University, Nashville, TN 37235
[2]Institute of Polymer Production Technology, Department of Chemical Engineering, McMaster University, Hamilton, Ontario L8N 3Z5, Canada

A series of polyacrylamide-co-sodium acrylates were prepared through alkaline hydrolysis of well fractionated polyacrylamides. The polymers were subsequently analyzed by light scattering, viscometry and size exclusion chromatography in aqueous Na_2SO_4 solutions. The dilute solution properties such as the specific refractive index increment, intrinsic viscosity, Mark-Houwink parameters and elution volume were found to systematically depend on the ionic content of the copolymer. These relationships were used to develop quantitative techniques for the molecular weight characterization of polyelectrolytes, including the use of the universal calibration method.

The molecular weight characterization of polyelectrolytes is relatively difficult since variations in copolymer composition alter the electrostatic environment and hence the size and structure of the polymer chain. Hydrolyzed polyacrylamide is therefore often used for methods development since it has a random distribution of charged groups along the backbone, and the molecular weights of the parent polyacrylamide can be estimated accurately by conventional techniques.

Copolymers of acrylamide and sodium acrylate can be prepared by direct copolymerization or derivatization from polyacrylamides. When the ionogenic monomer is introduced through a free radical addition mechanism, long acrylate and amide sequences form, with the microstructure conforming to Bernouillian statistics *(1,2)*. By comparison, polymers produced through alkaline saponification of the amide side chain are atactic *(3)* and have a relatively uniform charge density distribution which maximizes the viscosity increase for a given molecular weight and improves the polymer's performance *(4)*.

The purpose of this research is to develop valid characterization methods for copolymers of acrylamide and sodium acrylate. Initially a series of polyacrylamides will be synthesized and fractionated to yield narrow molecular weight distributions. The fractions will first be measured in their nonionic form where light scattering, viscometry and size exclusion chromatography (SEC) are more reliable and accurate than for ion containing polymers. The polyacrylamides will then be hydrolyzed

0097–6156/94/0548–0162$06.00/0

to various degrees and analyzed in their ionic form. The measured weight average chain lengths will be used to evaluate the absolute accuracy of the polyelectrolyte characterization methods. The results of these studies will also be applied to SEC calibration. Therefore we have employed aqueous Na_2SO_4 solution as a solvent and chosen to fractionate over a Mw range between 1.4×10^4 to 1.2×10^6 daltons. A second objective of this work is to investigate the validity of the universal calibration for polyacrylamide-co-sodium acrylate (PAM-co-NaAc) and nonionic polyacrylamides. Since the PAM-co-NaAc's which were prepared had narrow molecular weight distributions and relatively homogeneous composition distribution, the heterogeneity in composition and molecular weight is not expected to be an important factor in assessing the validity of the universal calibration.

Theoretical Considerations

The universal calibration method is based on Einstein's viscosity theory for spherical particles:

$$[\eta] = 2.5 \, N_A \, (V_h/M) \qquad (1)$$

where $[\eta]$ is the intrinsic viscosity, V_h the hydrodynamic volume of the particles, M the particle molecular weight and N_A Avogadro's number. From equation (1) it is obvious that the product $[\eta]M$ is proportional to the hydrodynamic volume. Since SEC separates polymer molecules on the basis of their hydrodynamic volume, one would expect that the product $[\eta]M$ is a function of the elution volume (V_e). The universal calibration is often expressed in logarithmic form:

$$\log[\eta]M = f \, (V_e) \qquad (2)$$

Accordingly, all polymer molecules with the same hydrodynamic volume elute at the same time. For example, if we designate a nonionic polyacrylamide with the subscript '0' and a PAM-co-NaAc with a sodium acrylate composition of x% with the subscript 'x' then the following identity holds:

$$[\eta]_0M_0 = [\eta]_xM_x \qquad (3)$$

The calibration curve for a single polymer can often be represented by the following equation, over a considerable range in elution volumes by(5):

$$M(V_e) = D_1\exp(-D_2V_e) \qquad D_1, D_2 > 0 \qquad (4)$$

This can be expressed in a linear form:

$$\log_{10}M = D_1' - D_2' \, V_e \qquad (5)$$

Experimental Methods and Procedures

Polymer Preparation. Polyacrylamides (PAMs) were synthesized by aqueous free radical polymerizations using potassium persulfate (BDH Chemicals, 99% purity) as an initiator and ethanol mercapton (BDH Chemicals) as a chain transfer agent. The polymers were heterodisperse in molecular weight, with polydispersities between 2.0 and 2.5 as determined by Size Exclusion Chromatography.

The synthesis conditions did not lead to preliminary hydrolysis as was determined by the ^{13}C NMR spectra. The spectra were recorded for a 10 wt% D$_2$O solution at 125.76 MHz and 70.9°C in the Fourier transform mode with inverse gate decoupling. The pulse width was 6.8 ms with an acquisition time of 0.557s.

Size Exclusion Chromatography. SEC chromatograms were measured with a Varian 5000 liquid chromatograph, using Toya Soda columns (TSK 3000,5000,6000 PW) each of which had dimensions of 7.5 mm x 30 cm and were connected in series. These columns contained polyether gel packing with hydrophilic OH groups on the surface *(6).* One of the advantages of this gel was that the surface groups were neither charged nor highly polar, therefore the adsorption and electrostatic interactions were greatly reduced *(7).* An aqueous mobile phase of 0.2 M Na$_2$SO$_4$ (BDH Chemicals; ionic strength 0.6 M) was used at a flow rate of 1.0 mL/min. Polymer solutions, 0.1% w/v, were prepared from dry polymer and the mobile phase. 100 uL of polymer solution was injected in each analysis.   sodium azide (0.1 wt%, Aldrich) and Tergitol NPX (0.01% wt%, Union Carbide Corp.) were also added to the mobile phase to protect the column from microorganisms and reduce adsorption. The chromatograms were recorded on a Varian CDS 401 Data Station. Peak broadening correction was performed using standard methods *(5,8).*

Fractionation, Hydrolysis and Copolymer Characterization.   A complete discussion of the fractionation conditions, the saponification reaction and the characterization of polyacrylamide-co-sodium acrylate has been presented in two previous publications *(9,10).*

Polymer Viscometry. The intrinsic viscosities of polymers in 0.2 Na$_2$SO$_4$ were obtained from the quadratic form as well as the conventional form of Huggins equation by the l·ast squares technique *(11,12).* The efflux time was measured with a No.75 Cannon-Ubbelohde semi-micro dilution viscometer at 25 ± 0.05°C.

Dialysis. A cellulose dialysis membrane (Spectra/Por 6) with a molecular weight cutoff of 1000 was purchased from Spectrum Medical Industries, Inc. (Los Angeles, CA). A small pore size was selected to prevent oligomers from diffusing into the dialysate. Prior to use the membranes were conditioned in the dialyzing buffer (0.2 M Na$_2$SO$_4$) for one hour, and rinsed with distilled deionized water.

Polymer solutions were prepared at a concentration suitable for light scattering *(13).* One hundred mL of these solutions were pipetted into the membrane which was sealed at one end with a dialysis tubing enclosure. After the second end of the membrane was secured, it was submersed in three liters of saline solution. This solution was housed in a four liter polyethylene container, isolated from the atmosphere.

After chemical potential equilibrium was obtained (120 h), the vessel was opened and the membrane removed. Several concentrations of the polymer were prepared by diluting the dialyzed solution with the dialysate. The samples were immediately analyᴚed by light scattering or differential refractometry.

Light Scattering. The molecular weight of each fraction was measured using a Chromatix KMX-6 LALLS photometer, with a cell length of 15 mm and a field stop of 0.2. These parameters corresponded to an average scattering angle of 4.8°. A 0.45 µm cellulose-acetate-nitrate filter (Millipore) was used for the polymer solutions. A 0.22 µm filter of the same type was used to clarify the solvent. Distilled deionized water with 0.02 Na_2SO_4 (BDH Chemicals, analytical grade) was used as the solvent.

For characterization of dilute solution properties of polyacrylamide-co-sodium acrylate, Wu *(9)* has determined that polyelectrolyte interactions are suppressed for Na_2SO_4 concentrations of 0.2 mol/L. This salt concentration was consequently selected as a solvent for the light scattering and SEC characterizations performed in this investigation.

The refractive index increment of the polymer in solution was determined using a Chromatix KMX-16 laser differential refractometer at 25°C, and a wavelength of 632.8 nm. The dn/dc was found to be 0.1869 for acrylamide homopolymers.

Results and Discussion

I. Characterization of Polyacrylamide. Eight PAM fractions with polydispersity indices 1.2 - 2.0 were obtained by fractionation. The homopolymers were subsequently analyzed by viscometry and light scattering with the intrinsic viscosities and weight-average molecular weights summarized in Table I.

The Mark-Houwink constants, K and a, were estimated with the Error-In-Variables method *(14,15)*. The variances in molecular weight were evaluated from Hunkeler and Hamielec's data *(13)*, while those in intrinsic viscosity were calculated from the present data using Chee's equation *(11)*. The variance in concentration was assumed to be negligible. The following equation has been established for PAM in 0.2 M Na_2SO_4 at 25±0.05°C:

$$[\eta] = 2.43 \times 10^{-4} \ M_W^{0.69} \tag{6}$$

The 95% confidence intervals for parameters K and a are:

$$K = 2.43 \times 10^{-4} \pm 0.36 \times 10^{-4}$$
$$a = 0.69 \pm 0.014$$

The 'a' value is consistent with that obtained by Kulicke *(16)* in 0.1 M Na_2SO_4 (0.7).

II. Light Scattering Characterization of Polyacrylamide-co-Sodium-Acrylate

Determination of the Refractive Index Increment at Constant Chemical Potential. The refractive index increment at constant chemical potential was determined as a function of the extent of hydrolysis *(10)*. The observed decrease in $(dn/dc)_\mu$ with acrylate content, a manifestation of the negative selective sorption of sodium sulphate, has been reported previously *(17-19)*. None of these authors have used the same solvent as in this work, and therefore direct comparison of the magnitude of dn/dc is not possible. Nonetheless, the validity of these measurements will be confirmed in the next section.

The equation:

$$dn/dc)_{\mu,PAM-NaAc} = 0.1869 \ (F_{NaAc})^{-0.076} \tag{7}$$

has been fit from this data so that the refractive index may be computed at any specific copolymer composition (F) between zero and thirty three percent sodium acrylate (NaAc).

Evaluation of the Molecular Weight Method for Polyacrylamide-co-Sodium Acrylate. Light scattering measurements were made for each of the five narrow polymer standards. The molecular weight data were normalized with respect to composition, with the corresponding chain lengths summarized in Table II. The polyelectrolyte chain lengths deviate on average by 7.68% from the original polyacrylamide homopolymer. Such an agreement is well within the random errors of aqueous light scattering (±10%)(12). We can conclude, therefore, that the measured molecular sizes of the ionic and nonionic polymers agree. It is worthwhile to note that without the correct refractive index increment, the estimation of the molecular weight of a thirty percent hydrolyzed polyacrylamide is 62% underpredicted. Therefore, the polyelectrolyte method, and the specific refractive index increments at constant chemical potential are accurate and reliable. Furthermore, it is sufficient to correct the optical constant for dn/dc)$_\mu$ in order to obtain accurate molecular weights of polyelectrolytes.

Table III lists the weight average molecular weight data directly obtained from preceding measurements of polyacrylamide-co-sodium acrylate as well as calculated from a the following stoichiometric conversion of light scattering data from nonionic polyacrylamides.

$$M_{ws} = M_{wPAM}/71.08(94.04x + 71.08(1-x)) \tag{8}$$

The values calculated from equation (8) will subsequently be used for the generation of calibration curves. Table III also lists the intrinsic viscosity and exclusion volume data for the polyacrylamide-co-sodium acrylates at each molecular weight and copolymer composition.

III. Viscometric Characterization of Polyacrylamide-co-Sodium Acrylate.

Mark-Houwink Equations for PAM-co-NaAc. In the Mark-Houwink equation the parameters 'K' and 'a' are constant only if the polymer composition, solvent and temperature are unchanged. For PAM-co-NaAc under the given conditions, 'K' and 'a' are functions of the hydrolysis degree. Klein et al. (20) observed a maximum value of exponent 'a' at about 40% sodium acrylate and a minimum value of 'K' at about 20% sodium acrylate. McCarthy et al. (21) showed some changes in the values of 'K' and 'a' with sodium acrylate but did not show definite trends. In the data presented herein 'K' is a decreasing function of the hydrolysis degree (HD) and 'a' an increasing function. This data is shown in Figure 1. In order to allow for interpolation of the Mark-Houwink equation for

Table I: Weight Average Molecular Weight (Mw), Polydispersity
(PDI),   and Intrinsic Viscosities $[\eta]$ of the Polyacrylamide
Fractions

Sample	F1	F2	F3	F4	F5	F6	F7	F8
Mw (daltons)	1.244E6	9,90E5	4.01E5	2.01E5	9.90E4	3.60E4	2.69E4	1.39E4
PDI	1.8	2.0	1.8	1.7	1.6	1.5	1.3	1.2
$[\eta]$ dl/g	3.804	3.555	1.733	1.096	0.6502	0.3220	0.2741	0.1754

Table II: A Comparison of the Polyelectrolyte and Nonionic
Molecular Weight Characterization Procedures

Weight Average Chain Length (r_w) of Polyacrylamide measured by the Nonionic Method	Weight Average Chain Length(r_w) of Polyacrylamide-co-Sodium Acrylate measured by the Polyelectrolyte Method		Percentage deviation from the Nonionic Characterization Method	
r_w (Nonionic Polyacrylamide)	r_w (10% hydrolysis)	r_w (30 % hydrolysis)	(10% Hydrolysis)	(30% Hydrolysis)
379	482	402	21.4	5.7
1,676	1,633	1,702	2.6	1.5
3,282	3,413	3,39.	3.8	4.6
6,563	6,412	5,959	2.4	10.1
13,859	13,845	10,466	0.1	24.6
			Average Deviation 7.68%	

Table III: Molecular Weight, Intrinsic Viscosity and Elution Volume of Polyacrylamide-co-Sodium Acrylates as a function of the copolymer composition

Sample Identification	Copolymer Composition (mol % sodium acrylate)	Weight Average Molecular Weight (Light Scattering) (daltons)	Weight Average Molecular Weight (Stoichiometric) (daltons)	Intrinsic Viscosity (dl/g)	Elution Volume (mL)
F7HY	10.77	35,800	27,800	0.3524	24.89
	36.00	31,700	30,000	0.4692	/
	36.08	/	30,000	0.4872	24.09
F5HY	10.30	119,600	102,300	0.8468	22.27
	16.70	/	104,300	1.049	22.83
	34.42	/	110,000	1.303	21.84
	39.20	135,500	111,500	1.342	21.66
F4HY	9.68	247,000	207,300	1.353	21.50
	31.90	/	221,700	2.200	21.09
	34.90	268,000	223,700	2.127	21.19
F3HY	6.40	/	409,300	1.961	20.30
	9.95	468,000	413,300	2.075	20.214
	14.90	/	420,300	2.640	20.02
	23.20	/	431,100	3.242	/
	26.00	/	434,700	3.470	19.73
	33.10	467,000	443,900	3.372	19.73
HY5*	9.46	/	1,061,500	3.979	18.33
	11.50	/	1,068,300	4.228	18.23
	20.15	/	1,097,000	5.329	17.81
	36.60	/	1,132,500	6.893	/
F2HY	10.30	1,014,000	1,029,000	4.354	18.38
	32.30	/	1,093,000	6.709	17.72
	33.40	822,000	1,096,000	7.067	17.72

* HY5 is an unfractionated polyacrylamide with a PDI of 2.5.

copolymer compositions not measured in this work the following polynomial expressions were regressed to the data:

$$a = C_0 + C_1 HD + C_2(HD^2) + C_3(HD^3) \tag{9}$$

where $\quad C_0 = 0.625 \pm 0.007$
$C_1 = 8.86 \times 10^{-3} \pm 1.27 \times 10^{-3}$
$C_2 = -2.405 \times 10^{-4} \pm 0.617 \times 10^{-4}$
$C_3 = 2.48 \times 10^{-6} \pm 0.89 \times 10^{-6}$

$$\log K = d_0 + d_1 HD + d_2(HD^2) + d_3(HD^3) \tag{10}$$

where $\quad d_0 = -3.36 \pm 0.024$
$d_1 = -2.39 \times 10^{-2} \pm 0.42 \times 10^{-2}$
$d_2 = 6.96 \times 10^{-4} \pm 2.05 \times 10^{-4}$
$d_3 = -7.37 \times 10^{-6} \pm 2.95 \times 10^{-6}$

The regressed equations (9 and 10) show a good fit with the Mark-Houwink parameters (Figure 1). Therefore, one can accurately calculate the values of K and a at any polymer composition of interest over the range 6 - 40% acrylate from these polynomials. In the absence of intrinsic viscosity data, equations (9 and 10) are useful for the construction of a universal calibration curve, particularly if on-line intrinsic viscosity detectors are not available. It is interesting to note that the sum (1000 K + a) is essentially constant (0.97-0.98) for copolymers of all compositions, and has a value of approximately unity.

IV. Size Exclusion Chromatography of Polyacrylamide-co-Sodium Acrylate.

Elution Volume and Molecular Weight of Polyacrylamide. Figure 2 is a plot of the logarithm of the molecular weight of nine nonionic polyacrylamide samples versus elution volume. A linear relationship is observed and the following equation has been regressed to the data ($\rho=0.998$):

$$\log_{10} M = 10.78 - 0.2505 V_e \tag{11}$$

In equation (11), M is the geometric mean molecular weight ($M = (M_n \times M_w)^{1/2}$).

Hydrodynamic Volume and Elution Volume of Polyacrylamide and Polyacrylamide-co-Sodium Acrylate. The hydrodynamic volume is plotted against the elution volume on a semi-log scale in Figure 3. The data for both ionic and nonionic polymers show no systematic deviation and the same correlation can be observed for all samples. Therefore, the universal calibration curve based on nonionic polyacrylamide seems to be a reasonable way to determine the molecular weights of polyacrylamide-co-sodium acrylate. The universal calibration has been linearly regressed to the data. A correlation coefficient of 0.995 was determined along with the following parameter values:

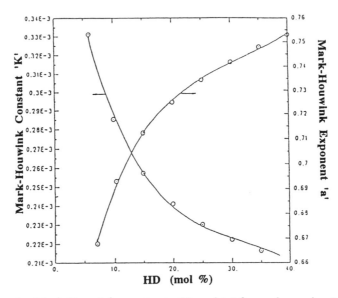

Figure 1: Mark-Houwink constants 'K' and 'a' for polyacrylamide-co-sodium acrylate at various hydrolysis degrees. (---): from correlated polynomials, (O): experimentally measured data point.

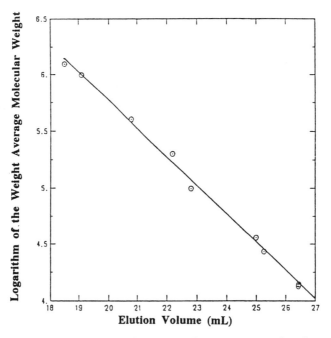

Figure 2: The logarithm of the weight average molecular weight (Mw) of nine nonionic polyacrylamide samples versus elution volume.

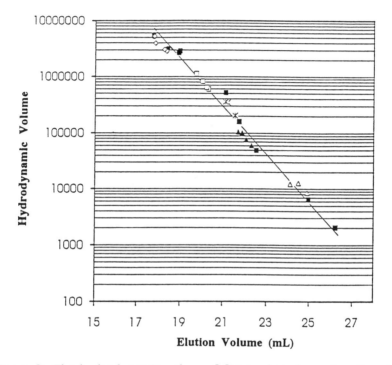

Figure 3: The hydrodynamic volume, [η]M, is plotted as a function of the elution volume (V_e) on a semi-log scale. (■): polyacrylamide homopolymer, (▲): F5HY, ([]): F3HY, (Δ): F7HY, (●): F2HY, (◊): HY5, (*): F4HY

$$\log_{10}M[\eta] = 14.50 - 0.4248V_e \qquad (12)$$

Effect of Charge Density on the Elution Volume. Figure 4 shows SEC chromatograms for polyacrylamide and polyacrylamide-co-sodium acrylate at various hydrolysis degrees. When the hydrolysis degree, or charge density, is increased the chromatogram shifts to lower elution volumes indicating a larger hydrodynamic volume. This is to be expected since an expansion of the polymer coil occurs when the electrostatic repulsion is increased.

Relationship between the Molecular Weight and the Elution Volume for Polyacrylamide-co-Sodium Acrylate. Figure 5 shows a plot of the logarithm of the molecular weight as a function of retention volume for various hydrolysis degrees. The parameters D_1' and D_2' have been regressed and are found to decrease linearly with the logarithm of the hydrolysis degree:

$$D_1' = 11.22 - 0.5055 \log HD \qquad (\rho=0.9985) \qquad (13)$$

$$D_2' = 0.2650 - 0.0148 \log HD \qquad (\rho=0.9994) \qquad (14)$$

The above correlations hold provided the hydrolysis degree exceeds 9%. For sodium acrylate levels less than 9% the D_1' and D_2' values for the nonionic polyacrylamide exceed those for the polyelectrolyte. This is a similar observation as reported in a prior publication (9) for the behavior of the Mark-Houwink exponent 'a' at low hydrolysis degrees. The above dependence can be explained through a combination of hydrogen bonding and electrostatic forces. For nonionic chains intramolecular hydrogen bonds between the C=O and NH_2 groups result in extended segments with an oriented structure (23). At low charge concentrations the oriented structure of the polyacrylamide is partly reduced due to a lower concentration of NH_2 groups. At the same time the concentration of COO^- groups is insufficient to generate sufficient electrostatic expansion to compensate for the lost expansion due to hydrogen bonding. Therefore, for sodium acrylate levels below 9% the chain expansion is less than for pure polyacrylamide.

Comparison of SEC Measurement with Light Scattering Data. Table IV shows the measured elution volumes and calculated weight average molecular weights for polyacrylamide-co-sodium acrylate over a range of copolymer compositions and molecular weights. The molecular weights were regressed with the use of the universal calibration (equation 12), as well as the 'linear' calibration curve (equation 5) where the parameters D_1' and D_2' are slight function of the copolymer composition (equations 13 and 14). Also tabulated are the weight average molecular weights measured by light scattering (equation 8). The linear calibration curve provided an average accuracy $(1-M_{w,SEC}/M_{w,LS})$ of $\pm 11.5\%$, whereas the universal calibration had an average accuracy of $\pm 19.8\%$. With neither calibration curve are the molecular weights systematically over or underpredicted. The superior data obtained with the linear calibration is not surprising since it is regressed to both molecular weight and copolymer composition data. The universal calibration has larger errors due to the

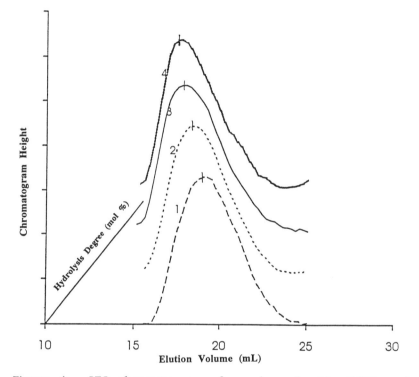

Figure 4:   SEC chromatograms for polyacrylamide (HY5) and polyacrylamide-co-sodium acrylate at various hydrolysis degrees. When the hydrolysis degree, or charge density, is increased the chromatogram shifts to lower elution volumes indicating a larger hydrodynamic volume. Curve 1: polyacrylamide homopolymer, curve 2: 11.5 mol% sodium acrylate, curve 3: 19.8 mol % sodium acrylate, curve 4: 30.8 mol % sodium acrylate.

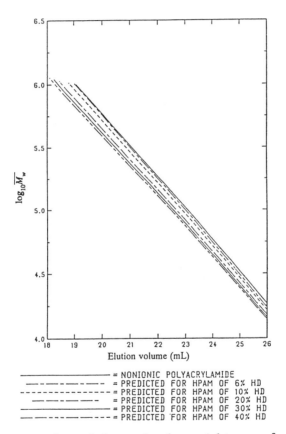

Figure 5: Logarithm of the molecular weight as a function of retention volume for various hydrolysis degrees. The calibration for polyacrylamide homopolymer is based on experimental data. The calibration curves for various hydrolysis degrees (HD) were calculated from equations (5,13,14).

Table IV: A Comparison of the Weight Average Molecular Weights as measured by Light Scattering with those estimated by Size Exclusion Chromatography

Sample Identification	Copolymer Composition (mol % sodium acrylate)	Weight Average Molecular Weight (Light Scattering) (Stoichiometric) (daltons)	Weight Average Molecular Weight (SEC, daltons) (Linear Calibration, Eqs. 5,13,14)	Weight Average Molecular Weight (SEC,daltons) (Universal Calibration, Eq. 12)
F7HY	10.77	27,800	27,300	23,970
	36.00	30,000	/	/
	36.08	30,000	34,210	38,330
F5HY	10.30	102,300	124,370	129,400
	16.70	104,300	128,110	132,090
	34.42	110,000	121,100	128,060
	39.20	111,500	129,940	148,280
F4HY	9.68	207,300	196,560	171,990
	31.90	221,700	187,370	157,960
	34.90	223,700	178,350	148,160
F3HY	6.40	409,300	435,410	383,780
	9.95	413,300	414,200	394,530
	14.90	420,300	413,500	374,890
	23.20	431,100	/	/
	26.00	434,700	422,460	378,760
	33.10	443,900	396,900	389,770
HY5*	9.46	1,061,500	1,230,500	1,299,100
	11.50	1,068,300	1,234,000	1348,200
	20.15	1,097,000	/	/
	36.60	1,132,500	/	/
F2HY	10.30	1,022,900	1,167,500	1,130,500
	32.30	1,093,000	1,229,100	1,399,200
	33.40	1,096,000	1,216,200	1,328,300

* HY5 is an unfractionated polyacrylamide with a PDI of 2.5.

variance introduced to the calculation from the intrinsic viscosity measurements, which lowered the correlation coefficient. Improved accuracy can be obtained with the universal calibration curve if the Mark-Houwink equation is employed with the constants 'K' and 'a' determined from the polynomials given in equations (9 and 10). Figure 6 provides visual verification of the agreement between the calculated molecular weights from the linear SEC calibration and those measured by light scattering

Conclusions

The universal calibration method is valid for the estimation of molecular weights of ionic polyacrylamide-co-sodium acrylates and nonionic polyacrylamides and can be based on a calibration from polyacrylamide homopolymers.

The molecular weights of ionogenic copolymers can be calculated from either the application of the universal calibration or the correlation $\log_{10}M = D_1' - D_2'V_e$, where D_1' and D_2' are functions of the hydrolysis degree (copolymer composition). The latter provides superior agreement with light scattering data.

The polyelectrolyte effects commonly observed in aqueous measurements of water soluble polymers are suppressed for polyacrylamide-co-sodium acrylate in 0.2 M Na_2SO_4.

Light scattering characterization of PAM-co-NaAc is valid provided the polymer solution and solvent are at chemical potential equilibrium. The equilibrium condition is primarily manifested in the specific refractive index increment, which must be estimated on well dialyzed samples to avoid introducing large systematic errors into the molecular weight estimation.

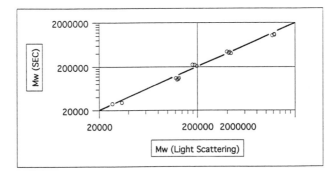

Figure 6: The weight average molecular weight determined by SEC ($M_{w,SEC}$) is compared with the weight average molecular weight determined by light scattering ($M_{w,LS}$). The molecular weights calculated from the SEC chromatograms and the universal calibration are in reasonable agreement with the absolute measurements as is evidenced by the proximity of the data to the 45° line. (0): Calculated from the universal calibration (equation 12).

Literature Cited

1. Troung,N.D., Galin,J.C., Francois,J., Pham,Q.T., *Polymer.* **1986**, *27*,459.
2. Candou,F., Zekhini,Z., Heatley,F., *Macromolecules.* **1986**, *19*,1895.
3. Troung,N.D., Galin,J.C., Francois,J., Pham,Q.T., *Polymer.* **1986**, *27*,467.
4. Ellwanger,T.E., Jaeger,D.A., Barden,E., *Polymer Bulletin.* **1980**, *3*,369.
5. Hamielec,A.E., Ray,W.H., *J.Appl.Poly.Sci.* **1969**, *13*,1317.
6. Hashimoto,T., Sasaki,H., Aiura,M., Kato,Y., *J.Polym.Sci.Polym.Phys.Ed.* **1978**, *16*,1789.
7. Hamielec, A.E., Syring, M., *Pure & Appl. Chem.* **1985**, *57(7)*,955.
8. Balke,S.T., Hamielec,A.E., *J.Appl.Poly.Sci.* **1969**,*13*,1381.
9. Wu,X.Y., Hunkeler,D., Pelton,R.H., Hamielec,A.E., Woods,D.R., *J.Appl. Polym.Sci.*, **1991**, *42*,2081.
10. Hunkeler,D., Wu, X.Y., Hamielec, A.E., *J.Appl.Polym.Sci.* **1992**, *46*,649.
11. Chee,K.K., *J.Appl.Polym.Sci.* **1985**, *30*,2607.
12. Fanood,M.H.R., George,M.H., *Polymer.* **1987**, *28*,2241.
13. Hunkeler,D., Hamielec,A.E., *J.Appl.Polym.Sci.* **1988**, *35*,1603.
14. Sutton,T.L., MacGregor,J.F., *Canadian J.Chem.Eng.* **1977**, *55*,609.
15. Reilly,P.M., Patino-Leal,H., *Technometrics.* **1981**, *23(3)*,221.
16. Kulicke,W.-M., Bose,N., *Polymer Bulletin.* **1982**, *7*,205.
17. Kulicke,W.-M., Horl,H.H., *Colloid and Polymer Sci.* **1985**, *263*,530.
18. Gunari,A.A., Gundiah,S., *Makromol.Chem.* **1981**, *182*,1.
19. Kulicke,W.-M., Siesler,H.W., *J.Polym.Sci.Polym.Phys.* **1982**, *20*,557.
20. Klein,J., Conrad,K.-D., *Makromol.Chem.* **1978**, *179*,1635.
21. McCarthy,K.J., Burkhardt,C.W., Parazak,D.P. *J.Appl.Polym.Sci.* **1987**, *33*,1699.
22. Brandup,J., Immergut,E.H., *Polymer Handbook*, 2nd Ed., John Wiley & Sons: New York, NY, 1975.
23. Kulicke,W.-M., Kniewske.R., Klein, *J.Prog.Poly.Sci.* **1982**, 403.

RECEIVED August 6, 1993